An Illustrated Histo[r]

MACK FIRE TRUCKS

1911-2005

Harvey Eckart

Iconografix

Iconografix
PO Box 446
Hudson, Wisconsin 54016 USA

Library of Congress Control Number: 2005927330

ISBN-13: 978-1-58388-157-6
ISBN10: 1-58388-157-3

05 06 07 08 09 10 6 5 4 3 2 1

Printed in China

Cover and book design by Dan Perry

Copyedited by Suzie Helberg

On the Cover:
Upper left: 1926 AC pumper, ex-Floral Park Centre, New York. Photo supplied by Greg R. Rhomberg
Upper right: 1948 Type 75 pumper, ex-New Holland, Pennsylvania.
Lower left: 1948 85LS pumper, ex-Selinsgrove, Pennsylvania.
Lower right: 1951 505A pumper Defender Fire Co. #3, Berwick, Pennsylvania.

Book Proposals

Iconografix is a publishing company specializing in books for transportation enthusiasts. We publish in a number of different areas, including Automobiles, Auto Racing, Buses, Construction Equipment, Emergency Equipment, Farming Equipment, Railroads & Trucks. The Iconografix imprint is constantly growing and expanding into new subject areas.

Authors, editors, and knowledgeable enthusiasts in the field of transportation history are invited to contact the Editorial Department at Iconografix, Inc., PO Box 446, Hudson, WI 54016.

Table of Contents

ABOUT THE AUTHOR

As a boy growing up in Berwick, Pennsylvania, I was fascinated with the sights and sounds of the Defender Fire Company's 1925 Mack AC pumper responding to alarms. I was especially thrilled to watch it pump for hours at major fires. At age 18, I joined this fire company and became a driver on the AC's replacement, a 1951 Mack 505A pumper. I was co-chairman of the committee to replace the 505A; and in 1978 a Mack CF611 pumper was purchased. I've been privileged to drive and pump Mack fire engines for over 50 years.

For more than 21 years I was the proud owner of a 1948 Mack Type 75 sedan cab pumper and a 1948 Mack L85 sedan cab pumper.

It has been my distinct honor and pleasure to author and edit six prior books on Mack fire apparatus. I am especially pleased to present this undertaking, which covers the complete history of Mack custom-built fire vehicles.

DEDICATION:

To the past and present employees of Mack Trucks Inc., the builders
To the world's firefighters, the users
To fire apparatus enthusiasts, the preservers

BIBLIOGRAPHY

Mack, by John B. Montville, Aztex Corporation, 1979
Bulldog, by John B. Montville, Aztex Corporation, 1979
Mack Fire Apparatus, A Pictorial History, by Harvey Eckart, The Engine House, 1990
Last of the Breed, by Harvey Eckart, The Engine House, 1994
Mack Model B Fire Trucks 1954-1966 Photo Archive, by Harvey Eckart, Iconografix Inc., 1997
Mack Model CF Fire Trucks 1967-1981 Photo Archive, by Harvey Eckart, Iconografix Inc., 1997
Mack Model L Fire Trucks 1940-1954 Photo Archive, by Harvey Eckart, Iconografix, 1998
Mack Model C Fire Trucks 1957-1967 Photo Archive, by Harvey Eckart, Iconografix Inc., 2000
Wheels of the Bravest 1865-1992, by John A. Calderone and Jack Lerch, Fire Apparatus Journal Publications Inc., 1993
The FDNY Super Pumper System, by John A. Calderone, Fire Apparatus Journal Publications Inc., 1985

ACKNOWLEDGMENTS

A project of this scope would not be possible without the input of many individuals who provide information, insight, encouragement, and enlightenment on the subject at hand.

I am especially indebted to the personnel of the Mack Trucks Historical Museum for their knowledge and support, namely Snowy Doe, Don Schumacher, Tom Loetzbeier, Dale Guth, and Rose Bundra.

Thanks to John Montville, fellow author and historian extraordinaire, for his encouragement, advice, and yes, even his chastisement at times.

Appreciation is also extended to all those whose photos are used, to those who provided information on specific vehicles, and to retired Mack fire apparatus salesman Bob Fosbenner.

Special thanks to my wife Hazel for her help and support, and for putting up with the many hours and miles spent in pursuit of the fire truck "bug" for these many years.

FOREWORD

I was pleased and honored to accept the invitation by Iconografix to author this work, which covers the entire period of custom fire truck production by Mack. It is indeed rewarding to share my passion for the Mack product in this manner with the many thousands of Mack fire truck enthusiasts.

It is not the purpose of this book to present a detailed history of the company, but to present an illustrated history on one specific product line. That is not to rely entirely on a "picture book" but to present enough historical details to supply a balanced perspective. For those readers who desire a more detailed history, particularly of the early period, I highly recommend the book *Mack* by John Montville, which is considered to be the "gold standard" of Mack history publications.

All major Mack custom fire apparatus models, except one, were based on a corresponding commercial truck series. I have chosen to present the various fire truck model progression based on the time period each series was produced.

In the black and white segment, factory photos were used, whenever available, because of their clarity and historical accuracy of what the vehicles looked like and how they were equipped at the time of manufacture. The color photo section reveals some of the same rigs when in service.

I strive for thoroughness and accuracy as much as possible within the constraints of some subject matter not thoroughly documented, and some "official" sources conveying conflicting information. I sometimes allow a personal observation to surface, which I trust will not be objectionable.

I am hopeful that you, the reader, will enjoy this book as much as I have enjoyed putting it together.

PREFACE

Mack Trucks Inc. – Corporate History:

The origin of the present Mack company goes back to 1893 when Gus and Jack Mack took over the Fallesen and Berry carriage wagon operation in Brooklyn, where they were employed. Brother William joined the venture in 1894 and the company operated as a partnership, simply called The Mack Brothers Company. In addition to wagon building, the brothers also offered repair services, which added to their growing list of accomplishments.

In 1900, Gus Mack had a discussion with Isaac Harris, which resulted in a contract to build a bus. This was the beginning of the present Mack company as a builder of motor vehicles. After delivery of the first Mack bus, a second order was placed by the same customer and Mack's reputation as a pioneer bus builder was secured.

The Mack Brothers Company was incorporated in New York in 1901. In 1905, the company was incorporated in Pennsylvania under a new name, The Mack Brothers Motor Car Company; and late in 1905 operations were being shifted to the Allentown, Pennsylvania, location. Brother Joseph joined the operation at that time.

Finances and competitive pressures resulted in a holding company being formed in 1911 called the International Motor Co. (IMC), involving the Mack Brothers Motor Car Company and the Saurer Motor Company. Hewitt Motor Company was added in 1912. IMC held responsibility for the sales and service function of the three truck builders.

In 1916, a new holding company, International Motor Truck Corporation, was formed and became the owner of IMC, which became the operating organization of Mack Brothers Motor Car Company Inc. In 1918, Mack became the sole survivor of IMC with Hewitt's demise in 1914 and Saurer in 1918, as truck builders.

The International Motor Truck Corporation's name was changed to Mack Trucks Incorporated in 1922, but International Motor Co. continued as the manufacturing arm of Mack Trucks Inc.

Jack Mack, who was Mack's president and the first vice president of IMC until his 1912 resignation, died in a car accident on March 14, 1924. In 1936, the name of International Motor Company (IMC) was changed to Mack Manufacturing Corporation.

The next major organizational change occurred in 1967 when Signal Gas and Oil Company, later known as The Signal Companies Inc., acquired Mack Trucks Inc.

In 1979 Renault of France bought a 10 percent share in Mack Trucks Inc. This was increased to 20 percent in 1982, and to 100 percent in 1990. In 2001, Renault sold Mack to Volvo, of Sweden, which made Volvo the world's second largest manufacturer of heavy-duty trucks.

Mack holds the title to the oldest continuous heavy-duty truck builder in America. In 1986 Mack delivered its one-millionth vehicle, a Super-Liner truck tractor.

Today, Mack is known primarily as a truck builder, but during its 100-plus year history, has produced an amazing variety of heavy vehicles and equipment. Besides trucks, Mack has built and sold:

- Buses and school bus chassis 1900-1960: 23,921
- Railcars 1905-1954: 93
- Locomotives 1921-1930: 20
- Truck trailers 1927-1944: 2,601
- Off-highway trucks 1961-1979: 2,672
- Custom fire apparatus 1911-1990: 10,508 (since 1937)

Throughout its 100-plus year history Mack has earned a reputation for performance that has made it one of the best known and most recognized companies in the world. "Built Like a Mack Truck" has become part of the language, denoting both strength and quality.

Mack's Role as a Fire Apparatus Manufacturer

The manufacture of fire apparatus is a small but highly specialized segment of the truck manufacturing industry. Motorization of fire apparatus in America is generally reported to have begun in 1906, and by the mid-1920s, most fire departments had completed the transition to all-motorized fleets. At the present time, there are approximately 5,000 fire trucks delivered annually.

The National Fire Protection Association (NFPA) adopted the first national specifications on municipal fire apparatus in 1914. Since that time, the National

Board of Fire Underwriters issued the standards for automotive fire apparatus from 1920 to 1948, when the present NFPA Committee on Fire Department Equipment was organized. There have been numerous revisions to the standard to keep it abreast of current practices. Various insurance rating and inspection bureaus, most of which are now part of the Insurance Services Office, witness acceptance tests of apparatus built under these specifications.

Fire apparatus falls into two general categories — commercial and custom. The definition of custom has changed slightly over time, but is generally considered to be a vehicle specifically designed, engineered, and built to be a fire apparatus. Many options are normally offered, and the manufacturer is equipped to meet specific, individual requirements, within reason. The commercial designation involves a regular commercial chassis, which is later modified, to some extent, for fire service. Custom units normally incorporate all requirements of the NFPA standards at the time of manufacture. Since the advent of diesel power, the distinction between commercial and custom has declined somewhat, with the primary difference now being in cab design and electrical components.

Mack apparatus has always been available as commercial chassis, but from the 1910s through 1990, Mack has also offered a complete line of custom chassis and complete custom units; built specifically for fire service. Since the early B series, Mack custom fire units have been identified by a number or letter combination in the serial number. This will be explained further in the individual chapters.

Up until the late 1950s, the major American fire apparatus builders were American LaFrance, Seagrave, Mack, and Ahrens-Fox. Ahrens-Fox bowed out in 1956 after suffering a slow, painful financial death. American LaFrance went out of business in 1985, but has been resurrected twice since then. Seagrave has been in continuous production of motorized apparatus since 1907, but is not among the current highest volume producers. Mack voluntarily withdrew from the custom fire apparatus industry in 1990, which will be expanded upon in Chapter 13. The current top producers are Pierce, Emergency One, and Kovatch Mobile Equipment; all of which rose to prominence in the last 30 years.

Mack was an anomaly in the custom fire apparatus field, as the only major player whose primary product was commercial trucks and buses, and not fire apparatus. This was an irritant to competitors, particularly to American LaFrance, who did take a stab at commercial truck production several times, with dismal success. Several of Mack's competitors in the truck field attempted to produce complete fire apparatus, but with limited success, after learning the manufacture and sale of fire apparatus involved an entirely different market and required considerably different selling techniques and support services. For over 50 years Mack was quite successful at maintaining separate engineering and sales divisions for their three major products: trucks, buses, and fire apparatus.

Mack's prominence in the custom fire apparatus field for nearly 80 years was the result of several factors:

1) From the very beginning Mack products were "over built," which resulted in high performance and dependability; characteristics very much appreciated in the fire service.

2) Mack was a vertically integrated manufacturer who specialized in designing and building their own major components, in stark contrast to the "assembler" builders who used components from various vendors to assemble a finished product. Mack highly touted their "balanced design," although recent years saw more and more of their product outsourced.

3) Mack gasoline and diesel engines were used in the overwhelming majority of custom fire vehicles and earned high marks for their performance and dependability. After the advent of almost total dominance of diesel power in the last few decades, Mack was the only manufacturer to build its own engines and drive trains.

4) Although the fire apparatus division had their own engineering department, total engineers at Mack exceeded those of their competitors and the fire apparatus division benefited from the accumulated experience through the production of heavy-duty vehicles of all types.

A fascinating aspect of custom fire apparatus production is that because of the low volumes and customization involved, a great deal of apparatus production deviates from the standard specification presented; which were obtained from standard sales literature.

Mack apparatus enjoys an exalted place in the apparatus enthusiast and collector fraternities. At many shows and musters there are frequently more Macks displayed than any of the other custom brands.

CHAPTER 1

The Early Years: 1911—1916

From 1900 to 1905 Mack's efforts were devoted to the production of 15 gasoline-powered buses at their New York City facility. Truck production began in Allentown, Pennsylvania, in 1905. Both trucks and buses carried the trade name "Manhattan." Conventional-type trucks with the engine and hood in front of the cab were called Type 1 trucks and cab over engine, or, as Mack referred to them, as seat-over-engine styles called Type 2 or 3. In 1909, smaller trucks of 1 to 2 ton capacities were introduced. These were all of the motor in front design and became known as the Junior line. The larger trucks were then referred to as the Senior line. Left-hand drive was standard on the Juniors and right-hand drive was standard on the Seniors. The Seniors had higher horsepowers than the Juniors but all were chain driven.

Numerous sources state that Mack delivered a tractor for a 75-foot aerial ladder to Allentown, Pennsylvania, in 1909, but I could find no confirmation of this in Mack chassis records or delivery lists. I am, therefore, showing 1911 as the first fire truck delivery by Mack to Cynwyd, Pennsylvania, now known as Bala Cynwyd. This was also Mack's first pumper, as a Goulds pump was fitted to a 3-ton special Senior fire chassis, and the

The first Mack pumper was delivered in late 1911 to the Union Fire Association of Lower Merion, located in a Philadelphia suburb now known as Bala-Cynwyd. It had a Senior chassis with a Goulds pump and Boyd bodywork.

body was furnished by James Boyd and Brother of Philadelphia, Pennsylvania. The chassis record only shows the engine as a large 4-cylinder (5x7). The paint is shown as Cochin red with gold stripe and fine white lines.

Early records do not distinguish between regular trucks and fire trucks, so an accurate figure on the number of fire trucks produced is not available. During the 1910s, the American fire service was rapidly converting to motorized equipment, and Mack, like many others, was supplying apparatus in increasing numbers.

Much of the apparatus delivered on the early Mack Junior and Senior lines were chemical engines and hose wagons. New York City received 96 Mack high pressure and chemical hose wagons between 1912 and 1916.

Many of the early Mack deliveries had Boyd bodies, but in 1914, Mack was also supplying their own. An interesting sidelight is that these early Mack 4-cylinder engines were equipped with dual ignition, as per an early brochure; presumably to enhance reliability on all trucks, not just on fire chassis. Also, during this early period, Mack claimed a "first" in the successful production of fire apparatus painted white, which did not turn yellow, as was the case with some others.

Over 3,100 early trucks were built between 1905 and 1916 with an unknown number being fire vehicles. The Senior and Junior lines overlapped the subsequent AB and AC models slightly, with some of the last sold as fire apparatus during 1916.

This Senior-chassis rig is listed as a three-ton combination chemical and hose wagon.

Although undated, this is believed to be an early Senior tractor-trailer aerial ladder. The hood is lettered AFD but the department is unknown.

Mack's first city service hook & ladder truck was delivered to Morristown, New Jersey. Some accounts list the delivery as 1910 but the records reveal a 1912 delivery. Although this was a Senior chassis, non-standard left-hand steering was specified.

Between 1912 and 1915, New York City bought 96 Mack hose and deluge wagons. This “fleet” shot shows six on display. These were Senior seat-over-engine designs, designated as a Type 3.

This conventional style chemical and hose wagon was delivered to FDNY in 1913. This was a Junior chassis with non-standard right-hand steering. Manpower didn’t appear to be a problem back then.

This 1913 Junior chemical rig shows Edia Fire Co. as the customer. It's posed in front of a classic fire station.

This Junior chemical wagon sports elaborate gold leaf striping.

This undated photo shows a Senior hose wagon on display at some type of show. The manifold on the running board indicates a Baltimore, Maryland, delivery.

This winter shot shows a Junior chemical and hose rig for Mount Holly, New Jersey. This rig featured a top-mounted bell and ladders on both sides.

This Senior certainly presents a racy appearance. The bodywork around the front seat and chemical tank highlights the versatility of the bodybuilder's art.

I'm not sure if this Senior tillered aerial is stuck or not, but there certainly is an audience observing whatever is going on.

This Senior chemical engine is believed to be a Boston, Massachusetts, rig.

These six FDNY ladder trucks are neatly posed. FDNY received twelve in 1915 and three in 1916.

This Junior chemical engine is believed to have been in service in the Huntington Manor Fire Department on Long Island, New York.

CHAPTER 2

The AB Fire Series: 1914—1930

The AB was Mack's first standardized high volume model. It was introduced in 1914 and over 51,000 trucks were built through 1937. Production figures on AB fire apparatus were not recorded separately. Fire apparatus was produced from 1914 until 1930.

AB models were first offered with worm drive rear axles but chain drive soon became standard. In 1920, double reduction drive replaced worm drive. In the early 1920s pneumatic tires became an option, and in 1923 a larger radiator became a noticeable change, as the radiator was no longer flush with the hood.

The AB and AC Mack fire apparatus models were both produced from the mid-1910s to the late 1920s. The AB was designed for light- and medium-duty applications. Early literature emphasized that both the AB and AC models used engines and chassis designed and manufactured solely for the purpose of fire service. AB models were offered with both chain and dual reduction drives. The AB was advertised as a moderately priced machine but emphasized it was developed from the motor truck, not the touring car.

One of the early AB deliveries was this AB1 chemical and hose car in 1915 to Bridgeport, Connecticut, with a high capacity deck pipe.

The following types of apparatus were offered on the AB models. At this point in time the type designations primarily reflected the function of the unit.

Type AB-1: Combination single chemical tank and hose car
Type AB-1A: Combination double chemical tank and hose car
Type AB-2: Combination pumping engine and hose car
Type AB-3: Triple combination pumping engine, single chemical tank and hose car
Type AB-3A: Triple combination pumping engine, double chemical tank and hose car
Type AB-4: Hose car
Type AB-5: City service hook and ladder truck
Type AB-5A: City service hook and ladder truck with single chemical tank
Type AB-5B: City service hook and ladder truck with double chemical tank
Type AB-6: Tractor
Type AB-9: Chemical engine
Type AB-10: Squad car

This AB3A with two chemical tanks and a pump was delivered to Darien (Connecticut) Fire Department in 1921. Solid tires and exposed chain drive were featured.

Chemical engines and hose cars comprised a large segment of AB production. When pumps were furnished, they were rotaries of 300-, 350-, or 400-gpm capacities.

The AB engine was Mack built with four cylinders and a bore and stroke of 4 x 5 inches. Displacement was 251 cubic inches. Brake horsepower started at 40 but was later bumped up to 60 when a larger 284-cubic-inch version was used.

AB models were outfitted with solid, pneumatic, or cushion tires. Standard paint was red ornamented with gold and black striping. Gold leaf lettering was used if requested. A bell and hand-operated siren was furnished. The standard Mack-built steel pumper body had a capacity of 1,000 feet of 2-1/2-inch hose.

The AB Mack fire apparatus served many towns and cities with dependable and economical fire protection, but was always overshadowed by the larger and overpowering AC model. Although the AB commercial production continued until 1937, fire apparatus production was supplanted in 1930 by the "early" B models with their sleeker appearances and six-cylinder engines.

This was a 1921 delivery of an AB3 to E. Rockaway, New York. A single chemical tank and pump are shown.

This 1921 AB3A was equipped with pneumatic tires, a pump, and two chemical tanks.

Keyport, New Jersey, received this 1921 AB5A city service ladder truck with chain drive and a single chemical tank. Considerable overhang of the ladders is evident.

The large life net is prominent on this AB5A ladder truck delivery to Farmingdale, Long Island, New York.

This 1925 City service ladder truck served Berwick, Pennsylvania, until 1947. The January 1925 photo shows a snowy setting.

These two 1925 AB3 pumpers are undergoing pump tests at the Little Lehigh River in Allentown, Pennsylvania.

This 1925 AB3 was a demonstrator equipped with pneumatic tires and shaft drive.

Lynbrook, New York, was a major Mack customer for many years. This AB1 had enclosed chain drive, chemical tank, and a deluge gun, but no pump.

This AB2 for Metuchen, New Jersey, was equipped with pneumatic tires and enclosed chain drive.

This long wheelbase quad was delivered to South River, New Jersey, in 1927.

This 1927 delivery to Hagerstown, Maryland, appears to be a chemical engine with longer than normal ladders, but not a full fledged hook and ladder rig.

This was a 1927 AB1 chemical and hose car for Indianapolis, Indiana, with pneumatic tires and shaft drive.

Delaware Twp., New Jersey, received this 1927 AB3A with dual chemical tanks, pneumatic tires, and enclosed chain drive.

Yorkville Hose in Pottsville, Pennsylvania, was well known for their Maxim quads, but this 1927 AB5A ladder truck preceded them.

East Syracuse, New York, was the recipient of this 1927 ladder truck with tremendous rear ladder overhang.

This 1927 fleet shot of the Merchantville (New Jersey) Fire Department shows an AP pumper, an AB chemical wagon, and an AB hook and ladder with life net.

A close-up shot of the Merchantville, New Jersey, chemical tank and booster hose arrangement. The dual intakes for the deck pipe would have to be fed by a pumper.

Similar to a fire truck, this 1930 Squad car for the NYC Police Department is shown in the Mack mid-town show room and would be one of the last AB emergency vehicle deliveries.

This undated photo shows an AB squad-type truck for the Boston Protective Department. That's a lot of truck for its limited function.

No date is listed for this AB1 single tank rig for Ridgefield Park, New Jersey. The tires and radiator design indicate an early AB offering. Note the graceful, flowing rear fender design.

No date is noted for this AB3 pumper for Maplewood, Missouri, which is shown drafting, presumably from the Mississippi River. There is no lack of observers.

Rochester, New York, is listed as the operator of this AB4 hose car. A minimum of warning and lighting devices is evident.

The AB four-cylinder engine is nicely shown in this 1925 delivery. The dual ignition consisting of both magneto and distributor is prominent.

CHAPTER 3

The AC/AP/AL Fire Series: 1916—1930

The AC Mack is probably the best-known truck in the world. Developed in 1915 and introduced in 1916, it epitomizes a heavy-duty workhorse that contributed to the development of the motor carrier industry and the industrialization of America early in the twentieth century. Over 40,000 AC trucks were built through 1938. A good number of these were fire apparatus, although separate production figures were not recorded. The AC fire apparatus line was phased out around 1928. The AC was the heavy-duty companion to the AB series and originally featured a 74-hp 4-cylinder engine, 3-speed transmission, hard rubber tires, and chain drive. Chain drive remained a hallmark of the AC to the very end.

Distinctive features of the AC were the blunt hood, cowl mounted radiator, and open "C"-style cab. The AC models achieved an immediate reputation for performance, stamina, and economy. Shortly after their debut, AC models were pressed into service for World War I demands. It was the British who started calling the AC models "bulldogs" because of their sturdiness and tenacious ability to keep going under adverse conditions. A secondary factor was the resemblance of the blunt AC hood to the snout of a bulldog, which was the symbol of Great Britain at the time.

The AC fire apparatus proved to be as overpowering as it was in commercial hauling. AC models were enthusiastically embraced by big cities as well as smaller fire departments. Initial offerings were 74-hp 4-cylinder engines of 471 cubic inch displacement, hard rubber tires, 3-speed transmissions, and chain drives.

Lakewood, New Jersey, received one of the very first AC chemical and hose wagons in 1916.

An early AC5 hook and ladder rig with solid tires and exposed chain drive.

Pneumatic tires were available in 1919, and by 1925 the horsepower was listed at 90 and the transmission now featured four speeds. Emphasis was placed on the point that each essential fire apparatus chassis part was specifically designed for the purpose of fire service.

AC fire apparatus catalogs listed the following types:

Type AC-1: Combination single chemical tank and hose car
Type AC-1A: Combination double chemical tank and hose car
Type AC-2: Combination pumping engine and hose car
Type AC-3: Triple combination pumping engine, single chemical tank and hose car
Type AC-3A: Triple combination pumping engine, double chemical tank and hose car
Type AC-4: Hose car
Type AC-5: City service hook and ladder truck
Type AC-5A: City service hook and ladder truck with single chemical tank
Type AC-6: Tractor
Type AC-7: Tractor drawn aerial hook and ladder truck
Type AC-8: Tractor drawn water tower
Type AC-10: Squad car
Type AC-11: Service and wrecking truck

In 1919, Mack introduced a Mack-designed pumper that used the AC chassis and an improved Northern rotary pump of 500- and 600-gpm capacities. Mack engineers designed this pump, but during the 1920s a Hale rotary pump was also offered. Mack claimed that their lateral intake and discharge features insured a smooth, steady flow, which resulted in "more gallons per horsepower."

A unique aspect of AC apparatus was the optional snowplow attachment for fire apparatus. An early advertisement stated, "By use of this plow, a fire fighting machine is converted to a fire protection unit as it opens up the streets so that alarms may be answered without delay." This feature no doubt appealed to city fathers more than to city employees.

A flurry of new developments in the late 1920s resulted in new combinations, and confusion in the model lineup. In 1926, an AL 6-cylinder engine was added to the AC line. Also in 1926 was the introduction of the AP 150-hp, 6-cylinder engine used in fire apparatus, also called the AP model. This resulted in a new "type" designation, which now involved primarily the engine size and pump capacity. The AP models were now designated as Type 15s, and 1927 advertising now listed this new type as available in either rotary or centrifugal designs, and in 750- or 1,000-gpm capacities.

An early AP delivery was a 1,000-gpm centrifugal pumper delivered to Seattle in 1927. The pump was a Byron Jackson, which was installed at their Berkeley,

This 1917 AC1 chemical and hose car for Little Falls, New York, features a top-mounted ladder rack in addition to the standard ladders. A switch from the mesh screen side hood shown, to louvers, occurred in late 1917.

California, plant. During 1927 two AP Type 15s were delivered to San Francisco equipped with Hale rotary pumps. The AP models were available with either chain or dual reduction drive. Four-wheel brakes, pneumatic tires, and electric starting and lighting were featured. Despite the longer hood to cover the 6-cylinder engine, the AP is frequently misidentified as an AC. New York City received the last AP models with bulldog hoods with the delivery of seven pumpers in 1930 and two rescues in 1931.

The AL bus and bus chassis was introduced in 1926 with the AL 6-cylinder engine being available in the AC model fire line. In 1927, the AL fire apparatus was introduced and a small number were built from 1927 through 1929. The AL was low slung with a bulldog-type hood and a front-mounted radiator. It had a 100-hp engine and a 750-gpm pump and was designated a Type 90.

The AC fire line offered tillered wooden aerials in lengths of 65, 75, and 85 feet. The hoist was a spring activated plunger type. In 1928, Mack designed the first engine-driven mechanically operated hoist for tractor-drawn aerial ladder trucks. This was considered a sensational development and enjoyed good acceptance through the mid 1930s. The first unit delivered with the new hoisting mechanism was delivered to Kearney, New Jersey, in 1928.

The AC fire apparatus enjoyed a production run of approximately 13 years but it, along with the companion AP and AL models, were phased out in the late 1920s when the demand for more speed and more modern styling swung heavily in favor of the "early" B series introduced in 1928. Relatively few AC fire trucks survived, as many were converted to other uses after retirement from fire service or were cut up for scrap during World War II. The ones that have been saved and restored command an inordinate amount of attention whenever displayed or demonstrated.

As might be expected, Allentown, Pennsylvania, has been a loyal user of the hometown product. Shown here are two pumpers and three chemical and hose wagons for the Liberty, Franklin, Goodwill, Hibernia, and Fairview fire stations.

This could be the start of the Indianapolis 500 race, but it's actually four AC-5As for delivery to FDNY in 1921.

The only difference between this winter shot in New York City and a current one would be the year and make of the rig.

ACs of this type (AC-11) performed various and sundry duties in large departments. This heavy-duty winch is on a five-ton, 168-inch wheelbase chassis.

This photo is undated but FDNY received six AC hose wagons in 1920 and another six in 1923.

Pneumatic tires became available in 1919 and this AC-3A triple combination pumper for Dover, New Jersey, is so equipped. It was delivered in 1923.

This 1924 photo contains no details but it is evident this AC tillered aerial has been engaged in a long and cold winter operation.

This AC3 pumper is the rig that got the author "hooked" on Mack fire trucks. It cost the Borough of Berwick, Pennsylvania, $9,328.13 in 1925 and was used in first line service until 1951. It had a 90-horsepower engine and a 600-gpm Northern rotary pump.

Huntington Manor, New York, received this 1926 AC-2 with enclosed chain drive. Tire chains were pretty much standard during winter conditions.

The first AC fire truck with an AL six-cylinder engine was this 1926 AC-3 for Phillipsburg, New Jersey. The pump was of 600-gpm capacity.

The AC-7 tractor drawn aerial hook and ladder was an imposing rig. This 1927 delivery features hard rubber tires, enclosed chain drive, searchlight, and wooden aerial ladder.

Neatly posed for a fleet shot are six AC-4 hose and deluge wagons, with exposed chain drive, for FDNY.

The regular AC fire apparatus engine was of 472 cubic inches with a bore and stroke of 5 x 6 inches. Horsepower was 90 with dual ignition and cylinder heads cast in pairs. Without muffler, this engine had a distinctive sound whether idling at 300 rpms or running full bore.

This view from the driver's position shows both the massiveness and simplicity of the AC cockpit. The lack of seatbelts and an enclosed cab would drive the modern "safety guys" crazy.

An early AP with the large 150-horsepower engine, pneumatic tires, and centrifugal pump was big news at this show.

Unlike the AC, the AP was available with shaft drive, and this is one of two so equipped for San Francisco, in 1927. The buses shown were also a major product line in this era.

Allentown plant 4A provides the background for this chain drive triple combination AP pumper, delivered to Rye, New York, in 1927.

With the hood removed, the six-cylinder engine is prominently displayed in this 1927 AP pumper.

The Spartan dash, siren, and searchlite are shown on this 1928 AP delivery to Ponca City, Oklahoma.

This is one of six AP pumpers delivered to St. Louis, Missouri, in 1928. Distinctive features are the pre-connected suction hose, disc wheels, and short rear body overhang.

The first AP aerial tractor was equipped with the shaft drive and delivered to Kearny, New Jersey, in 1928. Noteworthy is the combination of both pneumatic and solid tires on the same rig.

Only seven "bulldog style" pumpers were in service in FDNY. This 1930 AP 700-gpm pumper was one of them. A windshield and pneumatic tires were added at a later date.

Two AP rescue trucks were added to the FDNY roster in 1931. The rig assigned to Rescue One was rebuilt in 1940 and was in service until 1953, and then used as a spare until 1955.

This is a photo of a 1,000-gpm Hale "underslung" rotary pump as used on some AP pumpers.

This 1927 front-end shot shows the distinctive styling of the low volume AL fire model.

Low slung is the appropriate description of this 1927 pumper. A bench seat and single rear tires are shown.

Burbank, California, received this 1927 delivery which featured dual chemical tanks but no pump.

This "dolled up" white triple combination pumper with disc wheels was delivered to Pittsburg, Kansas, in 1928.

Pump and gauge arrangement on the Pittsburg, Kansas, rig is shown here.

The Humane Fire Company in Royersford, Pennsylvania, operated this rare 1928 AL ladder truck.

Ladders, pike poles, hose, and miscellaneous equipment dominate the rear section of the Royersford rig.

CHAPTER 4

The "Early" B Fire Series: 1928—1937

The need for more speed, comfort, and varying capacities led to the introduction in 1927 of the commercial B series, now referred to as the "early" B models to distinguish them from the B series introduced in 1953. In addition to meeting the above requirements, more modern styling was also evident. A total of 15,049 early B trucks were produced through 1941.

At this point Mack was now a major producer of custom fire apparatus. Fire vehicles based on the B series design were introduced in 1928 and produced through 1937 when the E series replaced them. Since fire apparatus production figures were still not recorded separately, the exact number built is not available. In 1930, a new serial number system was provided for identification of a fire unit. The first number was for the number of cylinders, the next two letters identified the chassis type, the next "6" denoted a fire vehicle, the next letter "C" or "S" referred to chain or shaft drive, and the chassis sequential serial number started at 1001. Thus, a serial number of 6BQ6S1001 translates to a 6-cylinder BQ chassis that is a fire vehicle, has a shaft drive, and is the first one built.

Sales brochures of this period reveal a total of 10 types offered:

Type	Chassis	Engine	Pump Size
19	BQ	150 HP (AP)	750 & 1000
21	BM, BX	225 HP (Hercules)	1000
50	BG	80 HP	500
505	BM	118 HP	500
55	BM, BX	100 HP	600
60	BM, BX	110 HP	600
70	BB, BC	100 HP	600
75	BX	110 HP	750
90	BK, BX	120 HP	750
95	BQ	130 HP	750 & 1000

Mack's cavernous plant 5C in Allentown is the setting for these "early B" fire trucks under construction in 1930.

All engines, except Type 21, were Mack 6-cylinder L-head designs with dual ignitions.

All pumps were rotary or centrifugal, except Type 21, which was centrifugal, and Type 50, which was rotary. Types 19, 90 and 95 were also available as aerial ladder tractors.

The B series offered various engines from 80 to 225 hp and pumps from 500 to 1,000 gpm in both rotary and centrifugal styles. Six cylinders, pneumatic tires, 4-wheel brakes, 4-speed transmissions, and shaft drive were now the norm. The Type 19 that used the big AP engine was "top dog" in the B lineup and it, along with the Type 21 were favorites of the big city departments. Mack claims to have pioneered the pressure/volume type of centrifugal pump used in the B series on the AP model back in 1927.

Mack advertising during the period boldly stated: "Mack offers only 6-cylinder engines in its fire apparatus as these simple, thrifty, highly standardized Mack-built engines accomplish everything necessary in Fire Department service both on the road and at the hydrant, and do this at lower speeds than engines with an excess number of smaller cylinders." (A not so subtle jab at the 8- and 12-cylinder engines of Seagrave and American LaFrance.)

During the 1930s several significant developments, including the following, occurred:

In 1932, a patent was issued for a bulldog radiator filler cap, and since that time, a chrome or gold bulldog has graced the hood or front of most Mack vehicles.

Air brakes were first used on Mack fire apparatus on the B series. Their use was limited, but most competitors did not offer air brakes, even as an option.

Practically all apparatus built up to and including the early 1930s was of the open cab configuration. Pirsch has the distinction of delivering the first completely enclosed custom built cab from a major manufacturer in 1928. Another milestone in cab design occurred in 1935 when Mack delivered America's first enclosed sedan cab to Charlotte, North Carolina. It was a Type 19 with a 750-gpm pump, a 100-gallon tank, and seating for 10 firefighters, which was a radical departure from the conventional two-man open cab designs. Seagrave later popularized the sedan cab with multiple deliveries to Detroit, Michigan.

Also in 1935, Mack delivered what was billed as the "largest fire engine in the world." It was a Type 95 with tandem axles, a closed three-man cab, a 2,500-gallon tank, and a 750-gpm pump. This was the forerunner to the modern pumper/tanker type vehicle. It was originally delivered to Lake, Wisconsin, which was annexed to Milwaukee. It is currently the centerpiece of a private museum in Wisconsin.

The Type 19 was the "heavy artillery" of the fire line in this era. This disc wheeled pumper demonstrator is a 1929 edition.

A 1930 Type 19 chassis showing the big AP engine and Byron Jackson centrifugal pump.

The engine-driven power hoist for tillered ladder trucks, introduced on the AP model, continued to be a popular rig into the late 1930s when metal aerials became the dominant types. Types 19 and 90 were the tractor models most used on B model aerials.

The early B series supplied the fire departments of both small towns and big cities for a 10-year period, bridging the gap between the powerful but slower AB, AC, AP, and AL units, and the newer, sleeker offerings of the E series to follow.

This 1931 Type 19 for Chester, Pennsylvania, shows that reel trucks are not a modern innovation.

This Type 19 is undergoing a pump test in March of 1935.

This 1935 Type 19 is historically significant as the first enclosed pumper in America. It was a Type 19 with a 750-gpm pump, 100-gallon tank, and 1,000 feet of 2 1/2-inch hose. After delivery to the Charlotte (North Carolina) Fire Department, it was involved in a collision in which a taxicab was demolished, but none of the firefighters were injured and the rig was retuned to quarters under its own power.

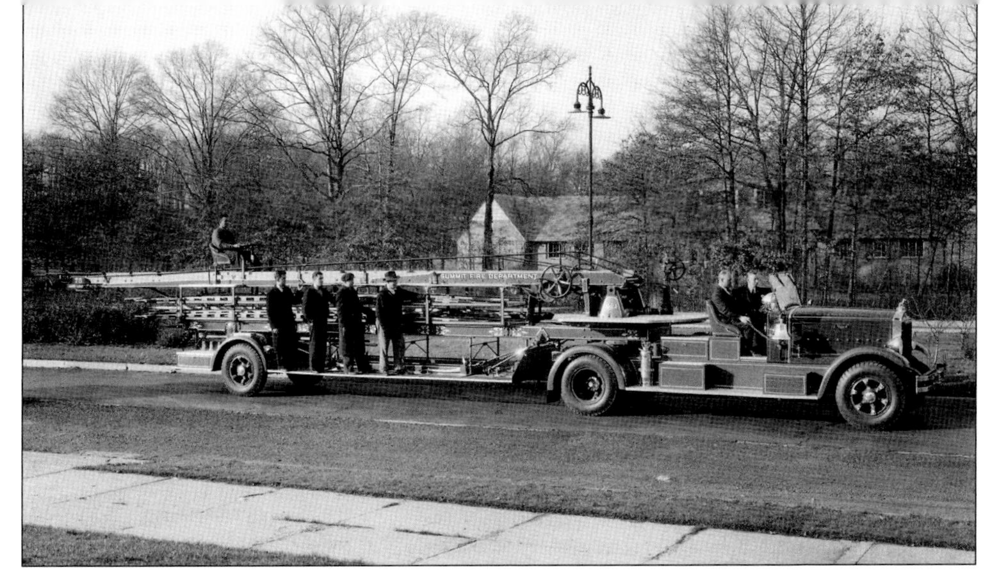

Summit, New Jersey, operated this 1935 Type 19 tractor with a Mack engine-driven aerial ladder hoist. It appears all the town fathers are checking the rig out.

One of twenty Type 21 1,000-gpm pumpers delivered to FDNY in 1936.

Crisp, clean lines mark the Type 21 chassis.

Nineteen more Type 21s were delivered to FDNY in 1937. Unlike the 1936 deliveries, these had closed cabs, and were also the first deliveries with the "New York subway straps" for the rear step riding members to hang on to.

The cutout doors were unique to the 1937 FDNY rigs. Also shown are the controls for the three-stage Hale pump and the deluge pipe handwheels.

This Type 50 rotary pumper lacks a windshield, hence the goggles on three of the four crew members.

A bus is shown in the background of this in-plant shot of a 1932 Type 50 pumper.

This 1932 Type 50 ladder truck shows the tremendous rear overhang of the wood ladders. The customer was A & M College of Texas.

The files yielded absolutely no details on this 1931 Type 55 tractor trailer tanker for Bogota, New Jersey. There is a good chance the specifications committee on this rig had a little too much to drink before their meeting.

Wilmington, North Carolina, received this Type 55 600-gpm pumper in 1936.

The Type 60 is represented by this spiffy 1935 quad for the Sanatoga Springs Fire Department.

This is a 1929 Type 70 single rear-tired pumper delivery for Auburn Fire Department No. 4.

This is a 1930 BC Type 70A hook and ladder chassis with an AB truck in the background.

Poughkeepsie, New York, received this Type 70 pumper in 1931. The low-mounted pump inlet was an asset in connecting hard suction hose.

Scranton, Pennsylvania, was not a major Mack customer, but they did receive two Type 75 BX pumpers in 1935.

Posed in front of a classic fire station is this 1936 Type 75 BX delivery to Revere, Massachusetts. Some 75s had hood doors while others had louvers.

This 1930 photo of a Type 90 pumper delivery to Mamaroneck, New York, is especially appealing because of the extremely neat display of the dozen pairs of boots and helmets.

The Type 90 tractor was a "big horse," as this 1931 aerial ladder for Montclair, New Jersey, illustrates.

It appears to be a bit of overkill as this Type 95 BQ for Hempstead, New York, was set up only as a hose and deluge wagon.

The Mack aerial of this era was a popular aerial ladder choice. Shown here is a 1934 Type 95 tillered aerial.

This 1935 Type 95 tractor was destined to pull an aerial ladder trailer in the service of the Bradford (Pennsylvania) Fire Department.

This "beauty of a brute" was billed as the largest fire engine in the world in 1935, for delivery to Lake, Wisconsin. It was a Type 95 with a 750-gpm pump, 2,500-gallon tank, three-man cab, 1,000 feet of 2 1/2-inch and 1 1/2-inch hose, and a 40-foot extension ladder. It is presently in a private museum in Wisconsin.

CHAPTER 5

The E Fire Series: 1937—1950

Introduced in 1936, the E series of commercial trucks were medium-duty trucks with streamlined styling, and were available in both conventional and cab-over configurations. In 1938, the Mack diesel engine was introduced and in 1949 an EH tractor-trailer toured the country to promote the diesel technology. The Mack diesel for fire apparatus service would not be available until 1960 in the B series. A total of 78,824 trucks were produced.

The E series fire apparatus were produced from 1937 through 1950. All except 22 were of conventional styles. Thirteen types, the most of any Mack fire apparatus series, were offered and a total of 1,596 were produced. A simple serial numbering system was used for custom E apparatus. It consisted of the type number, an "S" for single reduction drive, and then a four-digit sequential number beginning with 1001. As an example, 45S1002 would be the second Type 45 built with a single reduction rear axle. Sales brochures of the era offer the following details: (see chart)

E series apparatus was extremely varied and offered light-duty units of 100 gpm and 71 hp up to 1,500-gpm pumpers and quads with 225 horsepower.

The E series was the fist to use two brands of engines other than those of Mack manufacture. Types 25, 30, 40, and 45 were 500-gpm capacities. Continental engines were used in these models. At the opposite end of the scale, the Type 21s, which offered pump sizes

One of four Type 19s built was this 1940 open cab pumper for Schenectady, New York. The gold leaf striping and rear body curves contribute to a very handsome package.

from 1000- to 1,500-gpm, used Hercules engines. The Mack engines used were of both the older L-head design and the modern, overhead valve configuration that carried the "Thermodyne" nomenclature, which was acclaimed as the most advanced gasoline engine in use. All Mack engines featured dual ignition.

The braking system on most E models was hydraulic with vacuum booster, and the transmissions were of both four and five speeds. The Hale centrifugal pump was now standard, with the rotary as an option.

The open cab design of prior series' now gave way to a complete range of open, semi open, coupe, and sedan cab configurations of three door, four door, and canopy entries. In addition, special custom cabs and bodies were available as in the Deal, New Jersey, rig photograph, shown in the photo section.

The Type 80 extolled its virtues as compact, rugged, and powerful, a pocket battleship of 750-gallon range. Special note was made of the new Thermodyne engine, which was developed especially for fire service after eight years of research. Large gains in increased thermal and mechanical efficiency were claimed.

An early E model advertisement proclaimed: "Modernize with Mack," stating that a survey by Fire Engineering of 600 cities revealed that 50 percent of American apparatus were obsolete. The advertisement also stated that over 1,000 American communities operate Mack apparatus and also added, Mack has the facilities to build the finest – and does!

The 13 years during which the E series apparatus were built included the World War II years when Mack was authorized to build apparatus for cities that could get permission to buy, plus building units for the U.S. Government. One account lists 854 units delivered to the government during the war years.

Baltimore, Maryland, was the recipient of this massive Type 21 quad powered by a 935 cubic inch Hercules engine.

Type	6-Cylinder Engines	Pump Size	# Built
19	Mack Thermodyne 707ci 184 HP	1000	4
21	Hercules 935ci 225 HP	1000 to 1500	38
25	Continental 226ci 71 HP L-head	100-200	206
30	Continental 253ci 78 HP L-head	100-200	6
40	Continental 271ci 83 HP L-head	100-200	67
45U & 45	Continental 290ci 103 HP L-head	500	544
	Continental 330ci 123 HP L-head		
50	Mack 354ci 121 HP L-head	500	45
505U & 505	Mack 415ci 125 HP L-head	500	160
	Mack 510ci 161 HP Thermodyne		
55	Mack 468ci 140 HP L-head	600	52
60	Mack 524ci 150 HP L-head	600	6
70	Mack 431ci 124 HP L-head	750	30
75	Mack 525ci 150 HP L-head	750	393
	Mack 510ci 161 HP Thermodyne		
80	Mack 611ci 168 HP Thermodyne	750	45
			Total Built: 1,596

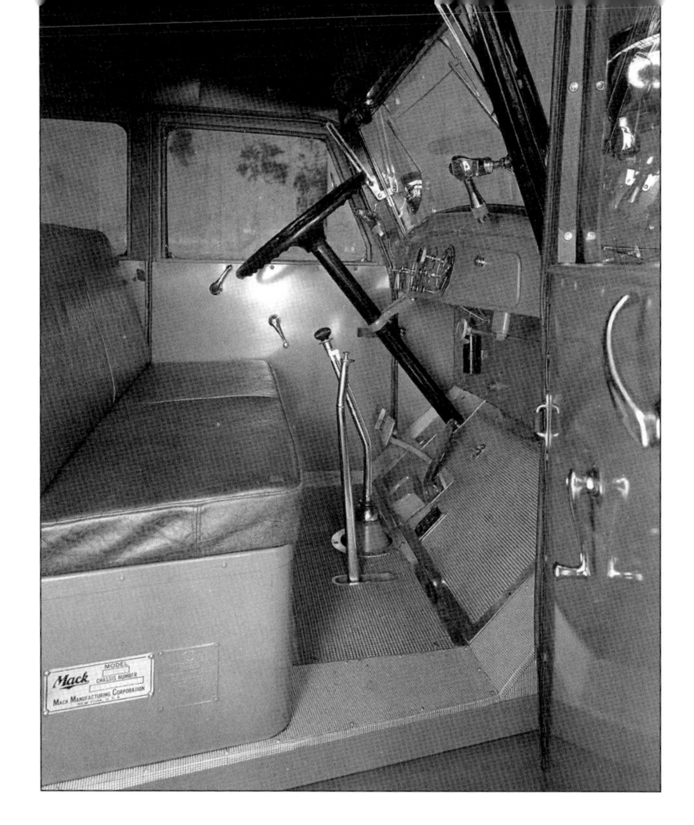

The front seat area of the Chicago sedan cab pumper shows the normal location of the serial number plate.

Type 21s were big city rigs and Chicago ran several 1,500-gpm sedan cab pumpers like this 1938 example. Shown also is a commercial E model moving van-type rig.

Unique to Chicago, two rear seats were provided. Except in extremely harsh weather, I'm told firefighters preferred to ride the rear step.

This closed cab 1939 Type 21 for Edgewater, New Jersey, is equipped with a Pirsch aerial ladder.

Only 206 of the tiny Type 25s were built. This 1941 closed cab pumper was equipped with an overhead ladder rack.

This open cab 1941 Type 30 was about as basic as a booster-type rig can be.

Mack didn't build many "bumper pumpers" but this 1938 delivery was the second Type 40 built and went to Cheltenham, Pennsylvania.

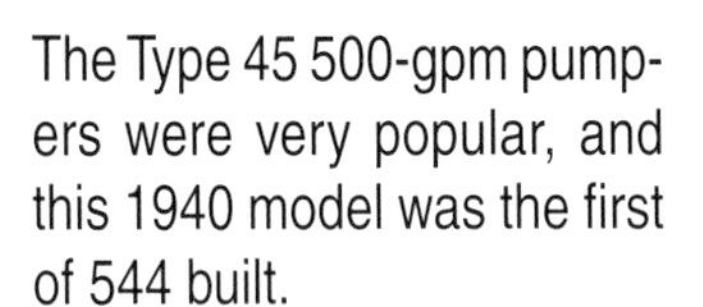

The Type 45 500-gpm pumpers were very popular, and this 1940 model was the first of 544 built.

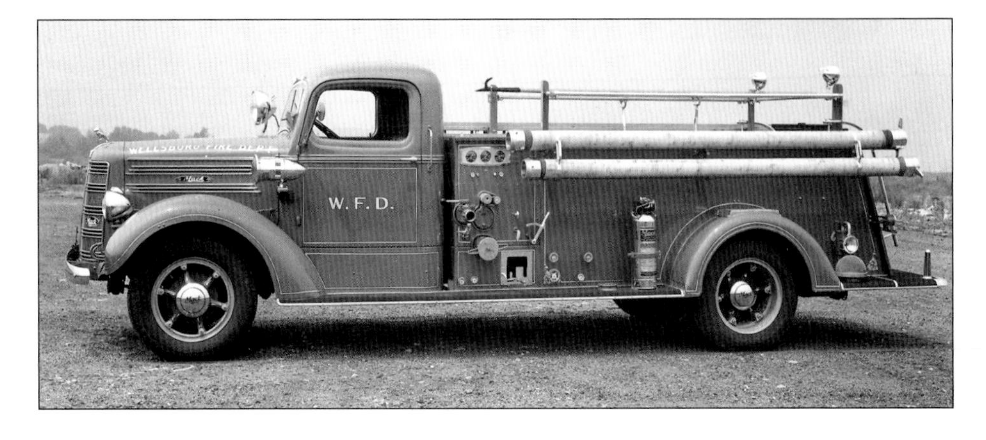

Wellsboro, Pennsylvania, is an American LaFrance town, but somehow this closed cab Type 45 pumper "snuck in" in 1941.

This semi cab Type 45 with a large overhead ladder rack was a wartime delivery in 1944 to Catasauqua, Pennsylvania. A photo of this rig, now owned by two collectors, appears in the color section in its present day form.

Smooth and sleek best describes this 1947 Type 45 closed sided city service ladder truck for Walden, New York.

Only seventeen of the cab-over Type 45U fire trucks were built. This is the third one built in 1940 for B. F. Goodrich Defense Plant, and is two-tone with a black roof.

Dubois, Pennsylvania, was a good Mack customer and this Type 55 pumper was delivered in 1941. The coupe cab with the small side window was frequently referred to as a streamlined cab.

Myerstown, Pennsylvania, was another good Mack customer, and in 1940 took delivery of this handsome Type 505 pumper.

This 1941 Type 505 sedan cab pumper was delivered to State College, Pennsylvania. They eventually ran both L and B model sedan cab pumpers also; and at one time the E, L, and B sedan cabs were all in service together.

Exemplifying true custom manufacturing, this 1941 Type 505 delivery to Deal, New Jersey, had what I believe to be a "one of a kind" cab design. The bid proposal referred to this as a "special 6-7 man canopy cab with side openings."

Mack was not a volume producer of squad or rescue-type rigs, but delivered this 1941 Type 505 to the Glenn L. Martin Airport in Nebraska.

City service ladder trucks were not rare but this 1948 Type 505 with a four-door cab was. It was delivered to Sioux City, Iowa.

Amityville, New York, received this very long and sleek Type 75 city service ladder truck in 1938. Both the seat and ladder compartments were of open design.

New Holland, Pennsylvania, ran this Type 75 canopy style sedan cab pumper from 1948 until 1981 when it was purchased by the author. An ENF510 Thermodyne engine powered the Hale ZD 750-gpm pump. *Bill Snyder photo*

Myerstown, Pennsylvania, still owns this 1948 sedan cab pumper. The disc wheels and flared roof warning siren and lights differ from the New Holland rig.

Open cabs were still built in substantial numbers after WWII, as was this 1947 Type 75 pumper delivered to Steelton, Pennsylvania. It is now privately owned and frequently seen on the muster circuit.

The Type 80 was introduced in 1938 and one of the first deliveries was this pumper for Baltimore, Maryland.

Simple clean styling highlights the cab area of the Baltimore rig. Wipers operating from the bottom of the windshield differs from the normal top mounting.

CHAPTER 6

The L Fire Series: 1940—1954

The heavy-duty L series commercial truck line was introduced in 1940 and quickly became a favorite of over-the-road drivers and their employers. The series offered Mack gasoline and diesel engines, as well as several vendor engines. The series consisted of conventional, cab-over engine, and off-highway configurations. Civilian production of the L series was constrained during the war years, but from 1945 to 1956 the L model trucks were familiar sights on the nation's highways. Over 31,000 L model highway trucks were produced from 1940 to 1956.

The L series fire apparatus was also introduced in 1940, and like their commercial brothers, became instant hits. Production totaled 1,453 units before production ended in 1954. Reminiscent of the 1920s, with the debut of the L in 1940, Mack was now offering two apparatus lines at the same time: the Es for light and medium duties and the Ls for the heavy jobs. After World War II, Mack held its own in spite of the increased competition from American LaFrance with their new cab forward line.

The serial number consisted of the type; model (L); single reduction axle (S), and sequential number — as in 85LS1001. The L model lineup was as follows: (see chart)

During the early part of World War II, Mack fire apparatus production was moved from Allentown, Pennsylvania, to Long Island City, New York. The chassis continued to be built in Allentown, with the bodies and final assembly being completed in Long Island City.

In 1948, the production lines of all manufacturers, including Mack, were turning out record numbers of apparatus to fulfill the urgent requirements of the fire service due to limited production during the war years. By 1948, these vehicles were once again works of beauty, with the return of chrome and gold leaf in generous amounts. The gleaming chrome radiator shell

FDNY Rescue Co. No 1 is known worldwide for both their workload and level of expertise. In 1948 Mack supplied a commercial LF chassis and Approved built the body for this famous workhorse.

of the L model Mack, along with the distinctive motor sound, made it instantly recognizable whether sitting in the station or working on the fire ground.

The L model apparatus production consisted mainly of pumpers and aerial units, although a complete line of city service ladder trucks, quads, quints, hose wagons, and specialized units were also available. Mack advertising proudly touted the fact that Mack built such vital parts as engines, clutches, transmissions, rear axles, brakes, steering gears, frames, sheet metal parts, and bodies in its own plants. Hale pumps were used primarily with Mack-designed and built pressure governors.

Vendor engines made by Hercules and Hall-Scott were installed in a small number of higher capacity units, but Mack Thermodyne engines of 611 or 707 cubic inch displacement powered the vast majority. The 707A and B dual ignition engine earned a solid reputation for high performance and dependability.

A few rotary gear pumps were installed, but Hale centrifugal pumps of two or four stages dominated. Sizes ranged from 750 to 2000 gpm, with 750 being the most popular. Five-speed transmissions were used initially, but later units used a standard four-speed direct unit. Hydraulic brakes were standard, but air brakes were specified in considerable numbers. Cab choices included open, semi open, closed, streamlined, and sedan cabs. The Long Island plant was renowned for the outstanding quality of their bodies. In 1951, fire apparatus production moved back to Allentown, Pennsylvania.

The L series quickly gained "king of the hill" status in fire departments across America including New York, Boston, Baltimore, Chicago, and Los Angeles. A notable accomplishment occurred in 1948 when Mack delivered two 2,000-gpm pumpers to Minneapolis, Minnesota: the first single engine pumpers of this capacity.

In 1940 Mack supplied FDNY with eight 19LS hose wagons. Unique items were the spare tire, single rear wheels, and a Buckeye exhaust whistle.

Also in 1948 a postage stamp was issued commemorating the 300th anniversary of the founding of the country's first fire brigade in 1648. From among more than 50 photographs submitted to the Post Office Department, an L85 Mack pumper from Riverdale, Maryland, was chosen to appear on the stamp.

Mack re-entered the aerial ladder market in 1949 with Maxim metal ladders of 65- and 75-foot lengths. This combination proved to be very popular with both large and small departments alike.

Revered by many as the best "top of the line" series ever built by Mack is the L model commercial and the fire apparatus series. They truly were classic vehicles in a classic era. Fortunately, many examples have been restored and preserved for future generations to enjoy. In a few cases, some are still answering fire calls today, the ultimate tribute to the L models' reputation for dependability. Many diehard L model enthusiasts firmly believe that, "on the eight day, God created the L model Mack."

Type	Engine (all 6-cylinder, gasoline)	Pump	# Built
19	Mack Thermodyne 707ci 225 HP	1000 (frequently 3 or 4 stage)	109
80	Mack Thermodyne 611ci 185 HP	750	141
85	Mack Thermodyne 707ci 225 HP	750	726
95	(same as 85)	1000	351
125	(same as 85)	1250	46
21	Hercules HXE 935ci 225 HP Hall-Scott 935ci 300 HP Hall-Scott 1091ci 324 HP	1000 to 2000	80
		Total Built:	1,453

Open cabs were still popular in 1943 when this 1943 19LS wartime delivery was made to Rockford, Illinois.

In 1946 Mack delivered this open cab 19LS tractor to Douglas MacArthur Fire Department in San Antonio, Texas. The trailer was an Ahrens-Fox and there was no bulldog gracing the radiator shell. The bell and siren were chrome but the headlights, windshield frames, and radiator shell were still painted items.

This 1946 21LS was a postwar experimental model. The body, door windows, and gauge panel differed from standard and this is believed to be a "one of a kind" offering.

In 1948 Minneapolis took delivery of two 21LS pumpers which were the first single engine 2,000-gpm pumpers delivered in the U.S. A giant 1,091-cubic-inch Hall Scott engine powered a Waterous pump which featured dual intake ports. These open cab brutes were used until 1974.

Sedan cabs were popular L model choices and this big 1,250-gpm 125LS was delivered to Green Bay, Wisconsin, in 1946.

Wauwatosa, Wisconsin, received this 125LS in 1948, which is drafting on a cold winter day. The three lights on the side of the cab were unique.

Dunn, North Carolina, received one of the first 80LS pumpers which were built from 1940 to 1944. Non-sealed beam headlights, a full windshield header, and rubber covered running boards were characteristics of very early units.

This solid white sedan cab 80LS pumper was delivered to Norfolk, Virginia, in 1941.

The chassis of the Norfolk rig clearly shows the 611-cubic-inch Thermodyne engine, 750-gpm pump, and Mack pressure governor.

This photo was used by Mack in a fire truck ad. It was a 1946 85LS painted a dark green for Martinsburg, West Virginia. One report states this rig pumped 1.6 million gallons of water from the basement of the Wheeling Hospital.

Although a rarity now, quads were a popular model in the L model era. This 1949 85LS for Oakhurst, New Jersey, featured a modified sedan cab arrangement.

I could only find two examples of open cab, crew cab pumpers. This 1950 85LS was delivered to Ridgefield, New Jersey. Note the pump panel is further back than normal.

The cab of the Ridgefield rig featured a full width rear bench seat. Tiffin, Ohio, also operated one of this design pumpers.

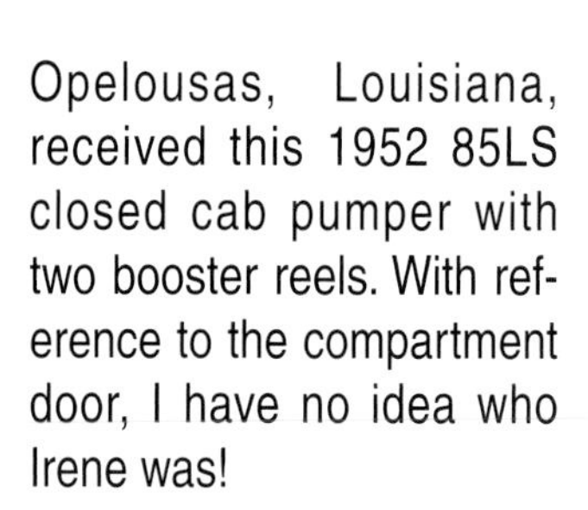

Opelousas, Louisiana, received this 1952 85LS closed cab pumper with two booster reels. With reference to the compartment door, I have no idea who Irene was!

A heavy-duty chassis was the basis for this 1953 85LS 75-foot aerial ladder truck. During this era, the Mack/Maxim combination proved to be a big seller.

It was an unusual occurrence to see a Mack chassis used on a competitor's model. I understand there was some chicanery involved regarding this 1954 85LS pumper with American Fire Apparatus body and pump, delivered to Talleyville, Delaware. This was the fifth from last 85LS built.

The LS 19, 85, 95 and 125 all used the Mack 707A or B engine which delivered powerful and dependable performance. These six cylinder engines were of 707-cubic-inch displacement and had 6 spark plugs on each side.

This 1951 95LS quad was delivered to St. Louis, Missouri. At this point semi cabs were the norm and extensive compartmentalization was rare.

Ten new 95LS pumpers in St. Paul, Minnesota, are engaged in an impressive public demonstration of their pumping capability in 1950.

Long Island, New York, was a major Mack stronghold. Elmont was the recipient of this classy 1952 95LS sedan cab pumper.

Unique features of the Elmont rig are the double coat and boot rails, seats over the hose bed, and combination stoplights and turn signals.

The Hale ZM centrifugal pump shown was common on L model pumpers.

The cockpit of this L model indicates hydraulic brakes. Dual starter buttons and a dash tachometer are also shown.

CHAPTER 7

The A Fire Series: 1950—1956

The A series commercial truck line was introduced in 1950 to replace the E series light- and medium-duty offerings and also to commemorate Mack's 50th year as a commercial vehicle builder. They were accordingly called the Golden Anniversary models. Over 20,000 trucks were produced through 1954.

Like their commercial counterparts, the A series fire apparatus replaced the E series. The bulk of the 191 fire trucks produced were built between 1950 and 1954, although several "specials" were built after the B series was in production. A model serial numbers begin with the type plus an A, and then the sequential four-digit number, as in 505A1005.

The A model was unique in several respects. Among the major Mack series, it covered the shortest production time period, and offered the least number of types. All of the engines used were of Mack manufacture and there were no A model aerial ladders.

Style-wise the A resembled the L model and is frequently misidentified as such. Both featured a chrome radiator shell but the smaller A cab had a single piece windshield whereas the L had a two-piece design.

Two L-head engines were used and a single overhead valve offering. The L-heads were called Magnadynes and the overhead valve designs were called Thermodynes. Both used dual ignitions. Five-speed transmissions and hydraulic brakes were standard, as was a Hale centrifugal pump of 500- or 750-gpm

The first 45A built was this 1951 pumper for White Salmon, Washington, equipped with an open cab and disc wheels.

The A model lineup was as follows:

Type	Engine (All 6-Cylinder Mack)	Pump	# Built
45A	ENF 331 331ci 130 HP L-head	500	44
405A	ENF 377 377ci 147 HP L-head	500	68
505A	ENF 510 Thermodyne 510ci 161 HP	500	36
75A	(same as 505A)	750	38
In addition, four 460As and one 475A were built.			5
		Total	191

A semi cab 45A pumper was sold to Sturgeon, Pennsylvania, in 1951. An extra section of suction hose was specified.

capacities. A push button primer was a $308 option. Mirrors, turn signals, and mufflers were not yet standard during this era.

All cab designs — open, semi open, coupe, and sedan — were available but few sedan cabs were delivered. Mack advertised the 45A and 405A as delivering maximum fire protection at minimum investment. A basic 405A pumper could be purchased for under $6,500. The 505A and 75A were advertised as the middleweight champs of the fire-fighting field. The advertisements also urged potential customers to compare not only specifications but also actual construction and performance.

A models are highly sought-after by collectors due to the low production figures and the "tough as nails" reputation. In addition, the A models were "pure Mack," as all major components except the pump were built in Mack factories.

This 1951 45A for Pittsburgh, Kansas, featured dual booster reels and a soft suction hose and tray.

Collingdale, Pennsylvania, received this 1951 45A with fender-mounted siren and a Roto Ray light, a popular item in the Philadelphia area.

A siren was missing on this 1951 45A pumper for Monroe, New York. Options included an extra hard suction hose, floodlights, and enclosed running board compartments.

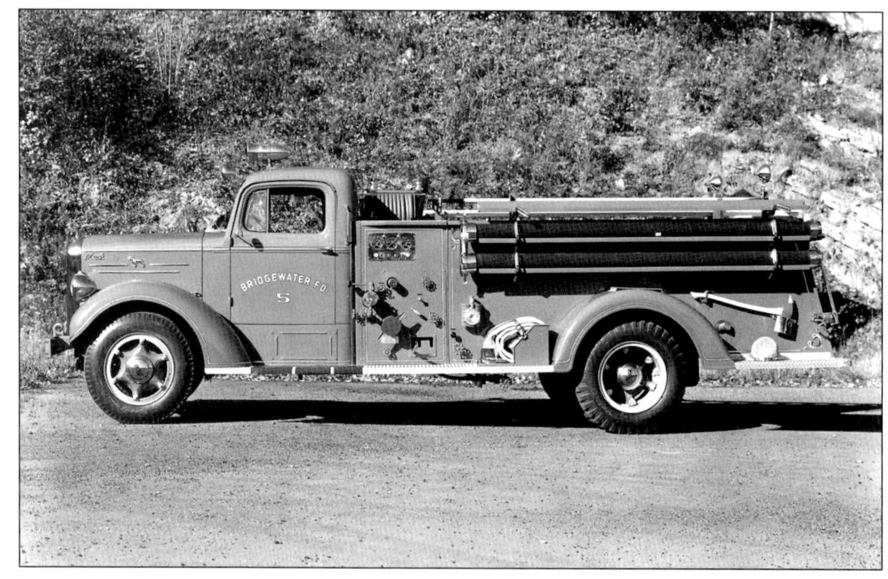

This 1952 closed cab 45A pumper for Bridgewater, Maine, includes a running board mount for the soft suction. Only 44 45As with the small 331-cubic-inch engine were sold.

This plain Jane open cab pumper for Modesto, California, was one of 68 405s built.

Protective wrapping was used on this 1952 405A pumper for export to Santiago, Chile.

Nothing real complicated about this driver compartment on the Santiago, Chile, pumper. The speedometer is calibrated in kilometers.

Aluminum ladders were specified on this 1952 405A city service ladder truck for Collingdale, Pennsylvania. This was one of the few A models equipped with air brakes.

Kaman Aircraft Co. in New Haven, Connecticut, ordered this 1953 405A pumper for their plant protection.

Cambridge, New York, received the first 505A built in 1951. At this point in time, large mirrors, turn signals, and mufflers were not standard items.

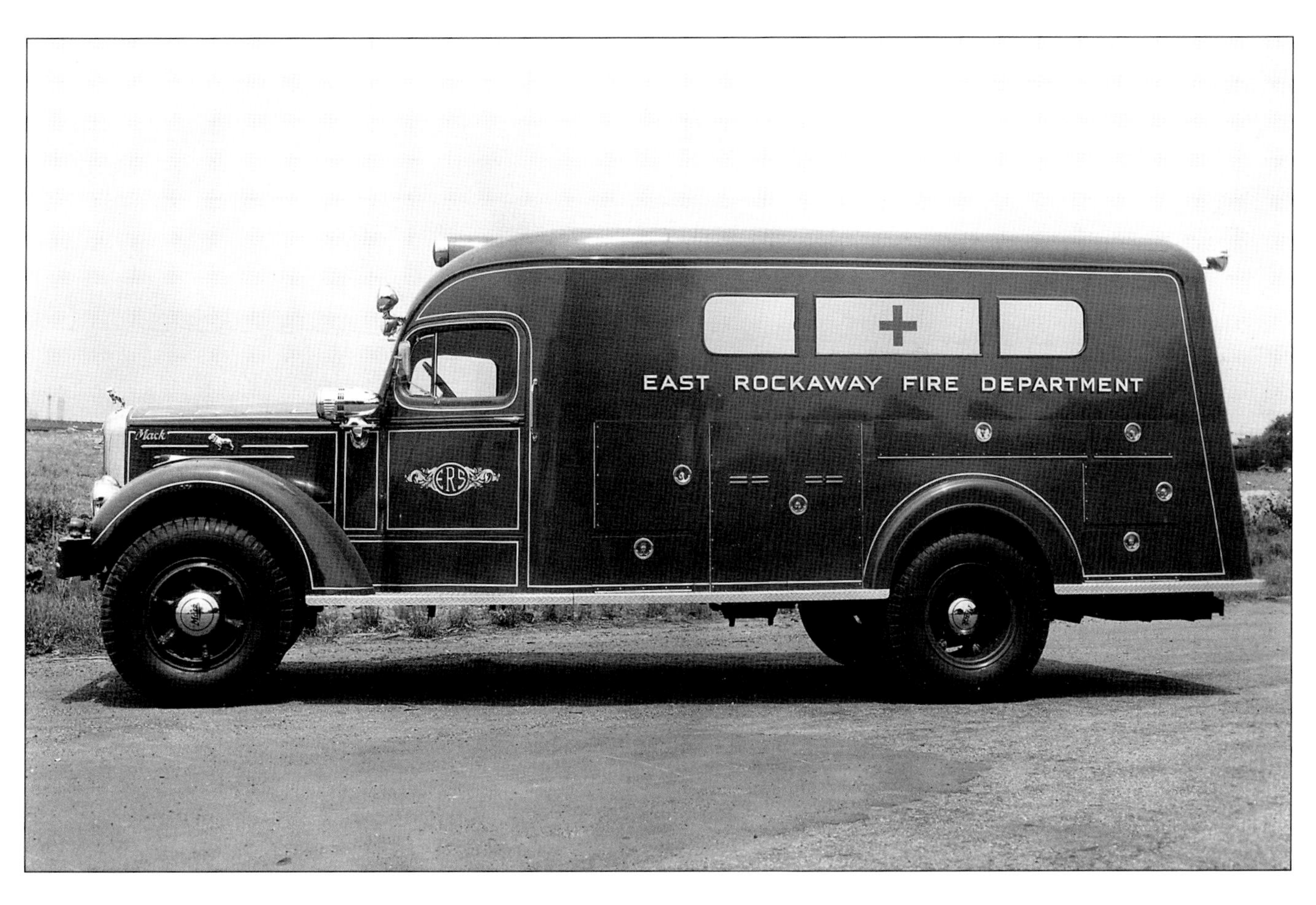

East Rockaway, New York, received the third 505A built in 1951, with a squad-type body.

This 1951 505A semi cab pumper was delivered to West Babylon, New York, in an area loaded with Mack-built apparatus.

The pump panel on the West Babylon pumper shows extra gauges, a gated 2 1/2-inch inlet, manual shift lever, and a Mack pressure governor.

Prospect Park, Pennsylvania, received this 505A city service ladder truck in 1953, powered by a Mack ENF510 Thermodyne gasoline engine.

Most of the rigs in this 1994 photo were collector owned and were present at the SPAAMFAA National Convention and Muster in Minneapolis, Minnesota. Models represented include "early" B, L, A, and C rigs.

Rockville, Maryland, is reported to have been the largest Mack user of all volunteer fire departments. This 1951 75A has a streamlined cab which was frequently referred to as a "Rockville cab."

It's not known why Mack mounted the bumper so high on the 505A and 75A models such as this 1951 75A pumper, but it detracted considerably to the otherwise attractive styling.

This 1952 75A pumper for Cooperstown, New York, featured a chrome bumper, turn signals, soft suction hose, and push-button priming.

Another industrial delivery is this 1952 75A open cab pumper for American Cyanamid Company.

Kaiser Steel received this 75A open cab pumper in 1953. As was common for industrial deliveries, no gold leaf striping was ordered.

Ormond Beach, Florida, received this 75A closed cab pumper in 1953. The standard 750-gpm Hale pump had two outlets on the left side, and one on the right.

A model sedan cabs were rare, but this pumper was on the roster in Akron, Pennsylvania. *William Stehman photo*

Restored AB pumpers are rare, but this one resides in Berlin, Maryland, and was photographed next to the Susquehanna River in Harrisburg, Pennsylvania. *Carl Moyer photo*

Also photographed in Harrisburg, Pennsylvania, was this AC pumper from Steelton, Pennsylvania, which is now privately owned. *Carl Moyer photo*

Stored in a barn for 31 years, this 1926 AC pumper was restored in the late 1990s and is now in a St. Louis, Missouri, museum.

This 1926 AC pumper was one of the many ACs built by the Baltimore, Maryland, shops and in service for many years. *Fire Museum of Maryland photo*

Kellog, Idaho, purchased this BQ Type 95 for $11,000.00 in 1934, and in the late 1980s it received a year and a half professional restoration.

Originally delivered to Hanover, Pennsylvania, this 1935 BX Type 75 was produced in 2004 as a superb 1:24 scale model by Yat Ming.

Salt Lake County, Utah, operated this 1935 Type 19 pumper which has been completely restored. *Jim Berry photo*

This fine example of a BG Type 50 pumper is collector owned in Richmond, Virginia. *Tom Herman photo*

New Holland, Pennsylvania, operated this 1948 E series Type 75 canopy sedan cab pumper for 33 years. The author owned this rig for 18 years.

Two private co-owners restored this Type 45 with overhead ladder rack from Catasauqua, Pennsylvania. The factory photo of this rig is shown in chapter five. *Ken Snyder photo*

This completely equipped 1938 quad is a remarkable restoration given the fact the New Jersey owner found it as a rusted hulk in a field. *John Floyd Jr. photo*

Berwick, Pennsylvania, operated this 1939 Type 505 brown and tan pumper for 35 years. It is presently on display at the Mack Trucks Historical Museum. *Glenn Edwards Studio photo*

This 85LS semi cab pumper served the borough of Hanover, Pennsylvania, and is now owned by a collector and apparatus restorer.

Green Tree, Pennsylvania, appropriately operated green apparatus. This 1950 closed cab quad presents a striking appearance with lots of chrome and striping.

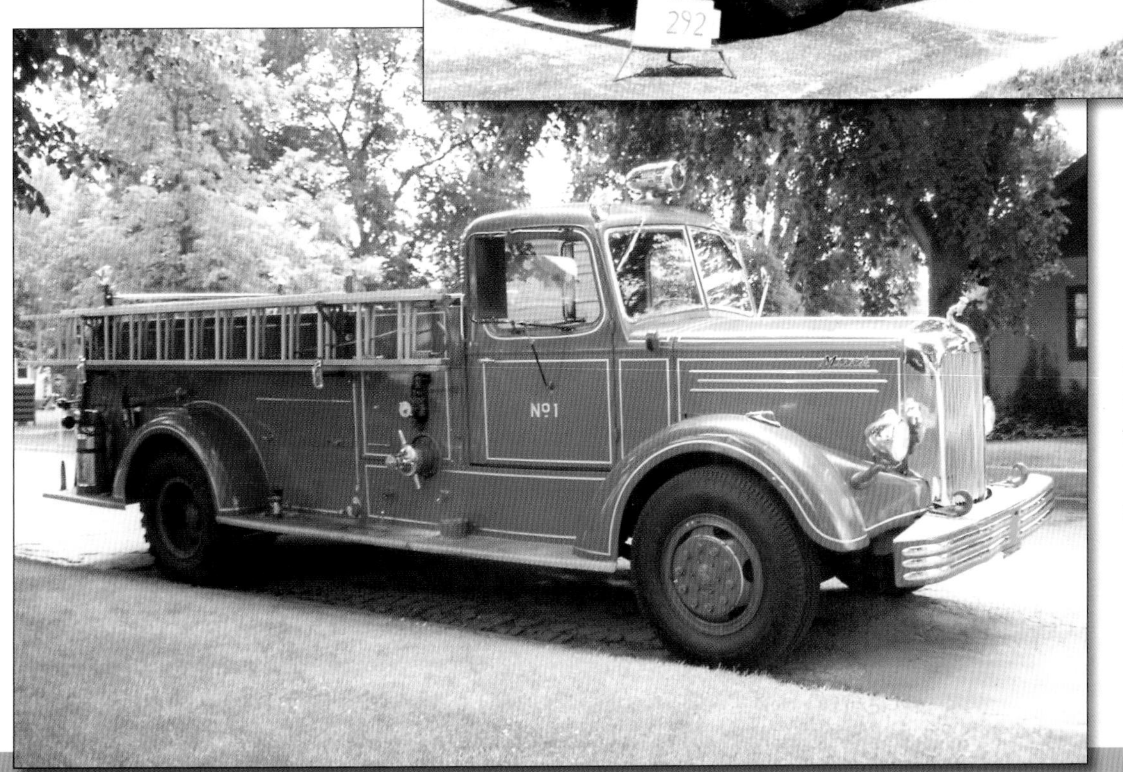

Litchfield, Minnesota, operated this big closed cab 1949 95LS pumper with disc wheels.

The author owned this 1948 85LS sedan cab pumper from 1987 to 2002. It cost $15,218.75 new and served Selingsgrove, Pennsylvania, for 39 years in first line service.

Berwick, Pennsylvania, operated this white 1951 505A pumper in first line service for 27 years and it is still on the roster as an antique.

This 1954 405A served the Utica, New York, area before being bought by a collector and fire magazine editor.

A model city service ladder trucks were rare such as this one delivered to Hellam, Pennsylvania.

This 1950 45A is distinguished by its unusual yellow and black color.

This 1958 B95 from the York, Pennsylvania, area is now collector owned and a frequent sight at parades and musters in northeast Pennsylvania. *William Stehman photo*

The front bumper, bell arrangement, and color scheme peg this as a 1960 B95 from the Chicago area. *William Stehman photo*

B model sedan cabs are highly prized by collectors, such as this 1961 B95 originally delivered to Bryan, Ohio. *Glenn Vincent photo*

"Jazzed up" best describes this 1965 B95 for West Haverstraw, New York. A stunning array of warning lights and devices, and turnout gear are shown.

One of six B21s delivered to Seattle, Washington, in 1958. *Bill Hattersley photo*

A two-tone paint job sets off this closed cab 1963 C95 pumper.

This collector owned C model pumper was displayed at the Mack 100th anniversary show in Lititz, Pennsylvania, in 2000.

Milton, Pennsylvania, operated this 1967 C95 which is now owned by a Pennsylvania State Police Fire Marshal; shown at the same show in Lititz, Pennsylvania.

This open cab pumper was one of the large fleet of C models operated by Harrisburg, Pennsylvania, area fire departments. *William Stehman photo*

The 1965 FDNY Super Pumper and Super Tender are shown when in service. *Joe Pinto photos*

The Howe plant where MB bodies were mounted is shown in the background of this 1975 MB pumper. *Roger Bjorge photo*

MBs ran for many years in Chicago as "flying snorkel squads."

This "plain Jane" tanker is a 1975 MB with Howe bodywork.

This 1967 R611F, for Port Vue, Pennsylvania, is unique because of its small 750-gpm pump, blue color, and semi open cab.

This basic R model pumper served in Carsonville, Pennsylvania. *William Stehman photo*

Canyon, Texas, operated this 1985 1,500-gpm R model pumper with Pierce bodywork.

This "foam monster" was built by Ward 79 on an R chassis with a 2,000-gpm pump and a 2,000-gallon tank.

This all-Mack CF fleet was in service in Reiffton, Pennsylvania. Apparatus shown are a 1975 CF685 Aerialscope, a 1973 CF608 pumper, a 1972 CF611 pumper, and a 1978 CF685 pumper.

Pomona, California, liked their Macks, including this rare 1967 open cab CF pumper.

Brown and tan were the unusual colors of this 1974 CF611 pumper for Berwick, Pennsylvania.

Lutherville, Maryland, apparatus was painted a distinctive black and white color combination such as this 1975 CF685 pumper, now collector owned.

Yonkers, New York, received the very last CF chassis built in 1990. It was a CF688FC fire chassis with an Interstate Truck Equipment Inc. body. *Tom Adams photo*

This is a very historically significant photo, which shows the last Mack bodied pumper, a CF688 model built in 1984; also the last custom MC chassis built in 1990, and an MC688 FC chassis with Saulsbury body. As of February 2005, both were still front line in Westbury, New York.

This is a rescue truck with the very rare MH chassis adapted for fire service.

Not all Mack fire vehicles were built in Mack factories!

Glendale, California, operated this big 1954 pumper on an over-the-road LT chassis.

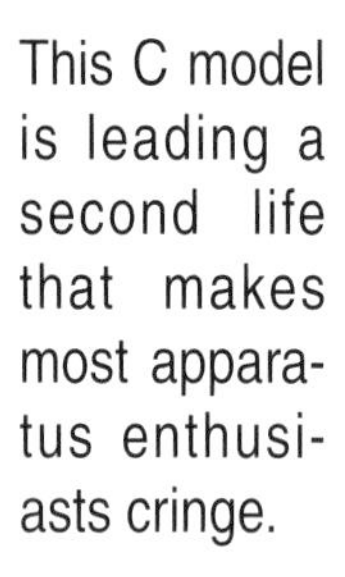
This C model is leading a second life that makes most apparatus enthusiasts cringe.

CHAPTER 8

The B Fire Series: 1954—1966

In 1953 there was a "new kid on the block" with the introduction of the B model, which would be one of Mack's most successful products ever. Well liked by drivers, its modern styling and advanced technology appealed to trucking companies everywhere, and it soon graced highways across the United States and Canada. By the end of production in 1966, B model commercial truck production totaled 127,786 units.

The fire apparatus version was introduced in the fall of 1954 and proved to be a worthy successor to the classic L and A series. Slightly less than 1,000 were produced during a 12-year production run. After 1957, the B fire apparatus shared the spotlight with the newly introduced cab forward C series.

The rugged good looks of the B apparatus, with its shiny radiator shell and dependable performance, made it a popular choice for fire departments when in service, and collectors and enthusiasts after retirement from active duty.

In addition to pumpers, the B Mack was available as a mid-mount and tractor-drawn Maxim aerial in 65-, 75-, 85-, and 100-foot lengths. Several Magirus aerials up to 144 feet long were also built. The B series was also widely used in tanker and rescue use. The B series was a wonderfully transitional series. It was available with open, semi open, coupe, deluxe, and sedan cabs; gasoline or diesel power; and hydraulic or air brakes. Because of higher speeds and longer runs, the closed cab became increasingly popular.

The B model was produced at the Allentown, Pennsylvania, plant until 1957, after which production was transferred to Sidney, Ohio. The Sidney plant was the former C. D. Beck Company plant, which Mack purchased in September 1956. The Beck plant produced Beck buses and Ahrens-Fox fire trucks prior to the Mack acquisition. Fire apparatus production was transferred back to Pennsylvania after the closure of the Beck plant in December 1958. Hahn Motors, of Hamburg, Pennsylvania, assembled Mack appara-

B21s were the big, bad boys of the B model lineup. Their giant Hall Scott engines and 1,250 and 1,500-gpm pumps put them in a class of their own. Memphis, Tennessee, received the first B21 built in 1957. The pre-connected suction hose arrangement was a Memphis feature for many years.

The only B21 open cabs were two for Ogden, Utah, in 1957. These 1,250-gpm brutes are now owned by a collector in New England, and attract a lot of attention when displayed.

tus from 1959 to mid 1961 before all production was again done at Mack's Allentown facilities.

When introduced, the B model standard specifications included 5-speed transmissions, hydraulic brakes, 200-gallon tanks, three open and three enclosed compartments, siren and horn (but not a bell), semi cab, and a soda acid and foam extinguisher. Before production ended the tank size would be increased to 300 gallons, all compartments would be enclosed, and the coupe cab and dry chemical extinguishers would be standard. In addition, an automatic transmission would be an option, as well as air horns and Federal Q sirens. Alternators replaced generators as standard items. In 1959, the factory list price of a B85 pumper was $17,200.

Two major developments occurred during the B and C model era. The first was a switch from Hale to Waterous pumps in 1960. I could find no official explanation for this, but conversations with individuals within the company indicate that Hale was slow or unwilling to grant a request of Mack. The result was the termination of an association that had lasted for more than 40 years.

The most significant development, however, was the sale in 1960 of two pumpers and a hose wagon to Bermuda, which were equipped with Mack ENDT673 diesel engines. Some accounts record this as the first sale of a diesel fire truck in the United States, but this is incorrect, as a Cummins diesel was installed in a New Stutz pumper in 1937 for delivery to Columbus, Indiana. The 1960 Macks were, however, the first delivery of a diesel made by the apparatus manufacturer. In contrast to the first delivery, which was regarded as an oddity, Mack's move provided the impetus for the eventual total dieselization of all heavy-duty apparatus. It took approximately 15 years for this to happen, and Mack delivered their last gasoline engine for apparatus use in 1973.

The vast majority of B models delivered with gasoline engines featured the premier 707C Thermodyne of 276 horsepower. Only nine B21s were built that used the "king kong" Hall-Scott engines that were nearing the end of the line. One was delivered to Memphis,

The B model lineup was as follows:

Type	Engine	Pump	# Built
21	Hall-Scott 935ci 300 HP	1250 & 1500	9
	Hall-Scott 1091ci 324 HP		
505	Mack ENF464 464ci 200 HP	500	48
	Mack ENF540 540ci 204 HP		
	Thermodyne gasoline		
75	(same as 505)	750	54
85	Mack ENF707C 707ci 276 HP	750	455
	Thermodyne gasoline		
	Mack ENDT673 672ci 230 HP		
	Thermodyne diesel		
95	(same as 85)	1000	236
125	(same as 85)	1250	53
405CF	Chrysler V-8 Hemi 354ci 204 HP	500	9
475CF	(same as 405CF)	750	42
(One B795F and one B20F were also built.)			2
			Total Built: 908

The Ogden, Utah, rigs' pump panel highlights the individual gauges, six discharge handles, and dual intake manifold. In addition to the Memphis and Ogden rigs, Seattle, Washington, received six B21s in 1958.

Tennessee, two were delivered to Ogden, Utah, and six were delivered to Seattle in 1957 and 1958. The Memphis and Ogden units used the 935 cubic inch engine and the Seattle units used the 1091cubic inch "Blockbuster."

Oddities were the 405CF and 475CF models. These were built to military specifications and used 354 cubic inch Chrysler Hemi V-8s. Forty-one were delivered to the U.S. Government and 10 to civilian users. The drive trains were smaller than standard and the entire unit was inferior to the standard Mack.

The commercial B model Mack is undoubtedly the most favored heavy-duty truck to be restored, and outnumbers other makes by a wide margin at truck shows. The B model fire apparatus is likewise the favorite of many enthusiasts and is widely acclaimed. Of all the apparatus designs, the sedan cab and straight frame aerials were especially handsome.

In spite of serious competition from within the company of the cab forward C model, as well as cab forward designs from competitors, the B model was an extremely popular and successful conventional-type apparatus. The R model that replaced it in 1966 was unable to sustain the volume and popularity of the B series.

Only 48 of the B505s, with their smaller engines were built. This big six-wheel pumper-tanker went to Princton Junction, New Jersey, in 1954.

The B series was renowned for their good looks. The chrome radiator shell was a nice styling feature.

This white pumper for Vernon, New York, was a 1956 B75 with dual booster reels and extra hard suction hose and tray. The Type 75 and 85 both had 750-gpm pumps but the 85 was favored because of its bigger engine.

This 1957 B75 quad for Middletown, New Jersey, was an impressive rig. Enclosed compartments were being built in increasing numbers.

The first of 455 B85s was this 1954 delivery to Pompton Plains, New Jersey. List price on a B85 pumper was $17,200.00

The cab of the first B85 indicates hydraulic brakes and a five-speed transmission.

This 1954 B85 quad for Jeanette, Pennsylvania, features a long wheelbase which reduces the rear ladder overhang.

Briarcliff Manor, New York, received this white B85 aerial in 1956. The dual windshield lights, spotlights, and twin sirens really sets this rig off.

Mack built a small number of aerials utilizing the Magirus rear-mount ladder. A good Mack customer, Milwaukee, Wisconsin, took delivery of this B85 in 1957.

A small conference appears to be in session regarding this Magirus ladder for Milwaukee. A bus production line parallels the fire line.

This 1957 delivery of a B85 sedan cab pumper features full side compartmentalization.

FDNY did not use any B model pumpers, but this B85 was delivered in 1959 with a Gerstenslager rescue body.

This is a classic semi cab B85 pulling a tillered trailer for the Reading (Pennsylvania) Fire Department. It was delivered in 1960.

Following previous deliveries of sedan cabs on the E and L series, State College, Pennsylvania, received this B95 sedan cab in 1966. Featured were dual air horns, front suction, and individual gauges. This rig is now collector owned.

This unusual photo was submitted by the Alpha Fire Co. of State College, Pennsylvania, and shows the B95 departing on a run back when rear step riding was the norm.

Baldwin, New York, ran this well equipped 1956 B95 semi cab pumper. Deck pipe, side-mounted booster, and turnout gear on rails are shown.

Fifty-three B125s were built including this 1957 delivery to El Monte, California.

True custom engineering marks the elaborate arrangement on the El Monte rig to accommodate a pre-connected rear suction hose.

This B125 pumper, along with another pumper and a hose wagon for Hamilton, Bermuda, made history in 1960. Mack switched from Hale to Waterous pumps but the big news was the installation of the first diesel engine manufactured by the apparatus builder. From this point on, diesel power would rapidly replace gasoline engines in heavy-duty apparatus use.

This famous publicity photo shows the Bermuda 1,250-gpm pumpers and hose wagon posed in front of the Queen of Bermuda ship. Although the diesel engines were revolutionary, the completely open cabs were rapidly becoming a "thing of the past."

Only nine B405CFs were built with Chrysler Hemi V-8 engines and 500-gpm pumps. This 1956 unit was built for stock and later sold to Rocky Mount, North Carolina, in 1957. The list price was $13,500. Note the Mack buses in the background.

The B475CF differed from the B405CF only in the pump size, 750 gpm. This is one of the 33 delivered to the U.S. government in 1956.

The B405CF and B475CFs were powered by Chrysler Hemi V-8s of 354-cubic-inch displacement.

CHAPTER 9

The C Fire Series: 1957—1967

In the fall of 1954 Mack introduced the B model conventional-type fire apparatus, which replaced both the popular A and L series. At this point in time Mack did not have a cab forward-type apparatus to compete with archrival American LaFrance, who introduced the post-World War II cab forward style to the fire service in 1945. Crown Body & Coach Corp. in California introduced their first custom fire apparatus in 1951, also utilizing the cab forward design exclusively. Ahrens-Fox, which was suffering a slow, painful financial death, was sold to General Truck Sales Corp. of Cincinnati, Ohio, in 1951. General Truck Sales Corp. contracted the actual construction of Ahrens-Fox apparatus to the C. D. Beck Company of Sidney, Ohio, in 1953. Beck was a small builder of buses. Renowned Ahrens-Fox salesman Frank Griesser, along with Beck engineers and designers, created a new Ahrens-Fox cab forward series in 1956 called the ECB and FCB models. Six of these were built by Beck after they acquired exclusive rights to manufacture all Ahrens-Fox apparatus in 1955.

In September 1956 Mack purchased the C. D. Beck Company for the purpose of building inter-city buses in anticipation of a large contract that never materialized. Mack engineers were impressed with the Ahrens-Fox cab forward design, and Mack purchased the production rights for this cab. This proved to be a favorable decision and in 1957 Mack began producing C model apparatus at the Sidney plant, as well as B model apparatus assembly transferred from the Pennsylvania facilities. The C model was the only major Mack fire apparatus line not based solely or in part on a commercial truck series. Mack also produced the inter-city bus model 97D (also based on a Beck design) at the Sidney plant.

Mack bus production quickly faded with only 26 units produced in 1958. The C model fire apparatus line was an entirely different story, as 1,055 were produced over the next 11 years, until they were eventually replaced by the CF series in late 1967. Fire apparatus production of both the B and C series was returned to Pennsylvania after closure of the Sidney

Only five C21s were built with the big Hall Scott 935-cubic-inch engines. All five went to Los Angeles, California, in 1958. All had disc wheels, 1,250-gpm pumps, and deck pipes.

plant in December 1958. Hahn Motors in Hamburg, Pennsylvania, assembled Mack apparatus from 1959 to mid 1961.

Although a latecomer in the cab forward market and not an original Mack design, the C model Mack was an instant success and became a major market contender. The C model design was one of the best-looking cab forward apparatus ever built.

Pumpers comprised the bulk of C model deliveries, but straight frame and tillered aerials in 65-, 75-, 85-, and 100-foot lengths were also popular. A significant development involving the C series was the introduction in 1964 of one of the first telescoping aerial platforms, called the Aerialscope, of 75-foot length. The C and CF Mack Aerialscopes were the almost-exclusive aerial platform-type apparatus in the New York City Fire Department for 30 years. A total of 13 Aerialscopes were delivered on C model chassis, with the first going to FDNY and five more following. They all had booms by Eaton/Truco, who had problems keeping up with production requirements.

Like the companion B model, the C model was a transitional series. It was available in both semi-open and closed canopy cabs. Both hydraulic and air brakes were offered, and in 1960 Waterous replaced Hale as the standard pump brand. Pump sizes ranged from 500 to 1,250 GPM with larger sizes available on a special order basis.

After 1960 Mack offered a diesel motor, which slowly but surely began to replace the gasoline powerplant in popularity. In an effort to convince reluctant potential customers, Mack conducted a 24-hour demonstration in Chicago in 1963. This was so successful that Mack then conducted a full week pumping marathon in Detroit, Michigan. A Mack C95 pumper pumped a continuous 1,000 gpm from noon of July 5, until noon of July 11, 1964, one full week without an engine shut-off. Mack proudly proclaimed this C95 pumped 10,160,640 gallons on only 1,108 gallons of fuel during this non-stop 7-day period. Mack credited this record to the durability of the ENDTF673 engine and the simplicity of the diesel engine, with less to go wrong. Mack stated the C95 consumed approximately 6.6 gallons of fuel per hour and that a comparable gasoline engine would have consumed about 15 gallons per hour. Not surprisingly, Mack diesels, as well as competitive makes, increased in popularity to the point where the gasoline engine is now only available and used in the smaller apparatus.

The small 500-gpm pumps were becoming extinct, and this was the only C505 pumper. Built in 1960, it had an ENF540 engine and was shipped to Cuba.

Today, over 38 years since C model production ended, some are still answering fire calls. A great many out-of-service rigs are now owned by collectors and fire companies who have restored them to like-new condition for future generations to enjoy. The C model is truly one of Mack's "crowning achievements" in the fire apparatus field.

The 1957 to 1967 period could be regarded as a "boom time" for Mack apparatus as the demand for both conventional and cab forward designs was met by the B and C model offerings. Both featured attractive styling, backed up by the performance of the 1960 introduction of diesel power, and the availability of the telescoping aerial platform in 1964.

The C model lineup included the following:

Type	Engine	Pump	# Built
C21	Hall-Scott 935ci 300 HP	1250	5
C505	Mack ENF464 464ci 200 HP	500	2
	Mack ENF 540 540ci 204 HP		
C75	(same as C505)	750	6
C85	Mack ENF707C 707ci 276 HP	750	308
	Thermodyne gasoline		
	Mack ENDTF673 672ci 230 HP		
	Thermodyne diesel		
C95	(same as C85)	1000	605
C125	(same as C85)	1259	129
		Total:	1055

Six C75s were built with a choice of two smaller Thermodyne engines and a 750-gpm pump. This New Jersey delivery was made in 1959. The Sidney, Ohio, plant is in the background.

The first two C85 pumpers went to Providence, Rhode Island, in 1957. They were closed canopy cabs with dual headlights and bumper-mounted turn signals.

Of the 52 C85s delivered to FDNY in 1958, this one was the only one delivered with a torque converter transmission. New York specs called for single headlights, single front spotlight, roof-mounted deluge gun, and booster reel mounted for rear deployment.

FDNY was on a C model buying spree in the fifties and this is one of two 1958 tillers with 100-foot Magirus aerials. There were also 24 C85 85-foot tillers with Maxim aerials delivered in 1959 and 1960.

Pumping into a deck-mounted deluge gun is this 1958 C85 in Larchmont, New York. A front suction is in use. Warning devices include a Federal Q and single air horn.

New Castle, Indiana, received this C85 in 1959. Turn signals and dual stop and tail lights were now standard items. The Federal Q siren was an extra cost option.

In 1961 two C85s were delivered to FDNY with 146-foot Magirus ladders. They were called High Ladders One and Two.

Farmingdale, New Jersey, received this C85 pumper in 1961. It was assembled at the Hahn plant in Hamburg, Pennsylvania.

Powered by a 707C Thermodyne gasoline engine and a Spicer 183 torque converter, this impressive C85 tillered aerial was delivered to Lynbrook, New York, in 1962.

The first of what would become an extensive fleet was this first Aerial-scope on a C85 chassis, delivered to FDNY in 1964. It had a 707C engine and a 75-foot Truco boom.

This C85 open cab quint was delivered to Connellsville, Pennsylvania, in 1962. It had a 65-foot Maxim aerial ladder.

The C95 was the most popular type with 605 built. This C95 pumper went to Tallman, New York, in 1966. The low-mounted warning light indicated a low clearance door was involved.

Like most West Coast deliveries, this C95 pumper for Aurora, Colorado, had disc wheels. Most departments opted for the chrome Mack West Coast style mirrors.

Jefferson City, Missouri, specified a C95 with a diesel engine, Federal Q siren, and bus style turn signals. It was delivered in 1966.

The department that received the first Mack pumper in 1911 received this open cab C95 pumper in 1966.

This white 1966 C95 diesel pumper blends in with the snow-covered parking lot.

Two-tone paint was not as common in 1967 when this C95 was delivered, as it is today.

This is one of 129 C125s built. Whittier, California, received this pumper in 1959.

This C125 was delivered to the Weyerhauser Co. in Portland, Oregon, in 1966.

Long Beach, California, was a loyal Mack customer and received this C125 pumper in 1966.

This was the C model assembly line at the Allentown, Pennsylvania, plant in 1964.

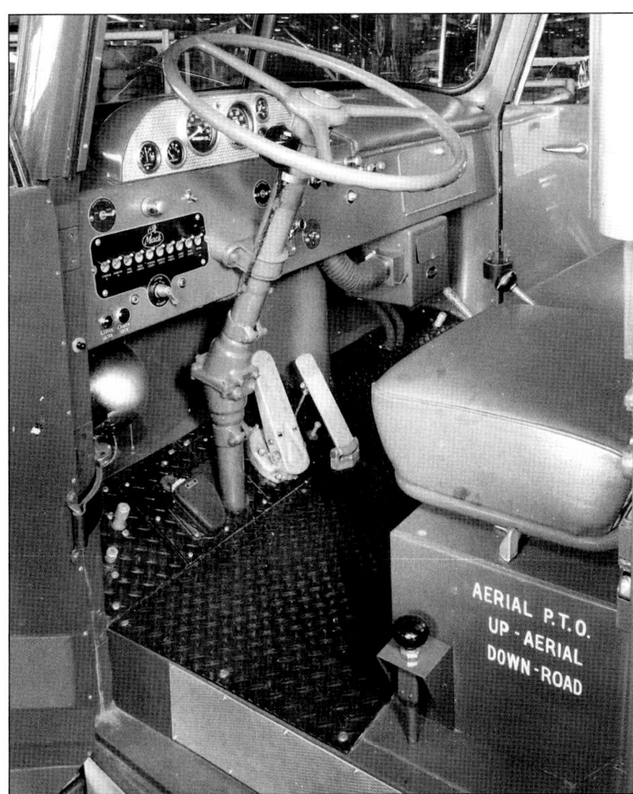

This C model cab area shows the optional dash-mounted tachometer, air brake treadle, and lack of a clutch pedal.

This C model chassis was equipped with a 707C gasoline engine, air brakes, and a Spicer torque converter.

CHAPTER 10

The Super Pumper System, 1965

The fire department of New York City, America's largest and busiest, faces big fires and in the early 1960s decided to fight back with big, new and bold weapons. The Super Pumper System, which was described as a fireboat on wheels, certainly fit the description. Mack accepted the challenge to build the system, which was not like anything produced prior to or since then. It's not known if Mack made any profit, but the publicity was certainly considerable. Their term "super pumper" has been brandished about frequently but never approached the mystique of THE true Super Pumper by Mack.

The Super Pumper System was conceived by W. Francis Gibbs, of Gibbs & Cox, an internationally known naval architect and marine engineer, and endorsed by the New York City Fire Department. On December 3, 1963, a contract was signed which authorized Mack Trucks, Inc. to build the Super Pumper, and accompanying hose tender, at a cost of $875,000. The purpose of the Super Pumper was to create a "land based fireboat" powerful enough to douse any holocaust, and to draft water from static sources when conventional water supplies were inadequate.

The Super Pumper System comprised of five pieces of equipment as follows:

The Super Pumper – A Mack F715FSTP tandem-axle tractor with an END864 V-8 diesel and Allison CLT4460 6-speed automatic transmission, pulling a tandem-axle trailer equipped with a DeLaval six-stage pump with a capacity of 8,800 gpm at 350 psi. The pump was driven by a Napier-Deltic T18-37C 18-cylinder diesel of 2,400 horsepower.

The Super Tender – A similar tractor to the Super Pumper with a huge, permanently mounted 10,000-gpm monitor, which could discharge a stream up to

The Super Pumper poses for its delivery photo in 1965.

600 feet. The trailer was equipped to transport 2,000 feet of 4 1/2-inch hose.

Three satellite tenders – Model C85F diesels with manual transmissions, each with a hose bed capacity for 2,000 feet of 4 1/2-inch hose and equipped with 4,000-gpm monitors.

The Super Pumper System was delivered in mid-1965 and its first fire was on August 12, 1965. The system operated all over the five boroughs of New York. It was used in many different configurations with all or part of the system responding as needed.

During its years of service, the Super Pumper and Tender responded to over 2,200 alarms, the last being on April 24, 1982. The system proved to be a versatile and powerful tool, but due to age and wear, it was taken out of service in 1982.

There was some talk of keeping the Super Pumper as a museum piece, but it was loaned to the U.S. Navy in Norfolk, Virginia, and then was eventually sold by bid. The current owner is a collector in Michigan. The original tractor has not been operable for some time but the Super Pumper was transported to the SPAAM-FAA show in Frankenmuth, Michigan, in 2004 via a Super-Liner Mack tractor. The appearance of the pumper has not changed much from when it was taken out of service.

A collector in Southern California presently owns the Super Tender. It is still pulled by the original tractor and is frequently displayed at parades and shows, including the 2004 ATHS show in Fontana, California.

In 1977, ComCoach rebuilt the three original C model satellite hose wagons. The front hubcaps were the only indication of their former lives. The current satellite fleet consists of six Mack MR chassis units and was part of the Maxi-Water System, which replaced the Super Pumper System. The satellites are currently paired with 2,000-gpm pumpers when their services are needed.

The Super Pumper System was truly unique, and apparatus enthusiasts are thankful both the pumper and tender are still in existence. Hopefully they can one day be displayed together, preferably in a museum setting. It is doubtful that Mack's accomplishment of building the world's largest municipal fire truck will be challenged or equaled anytime soon.

The DeLaval pump is shown. Above the wheel well are four of the eight 4 1/2-inch discharge outlets. At the rear are both 4 1/2-inch and 12-inch suction inlets.

This Napier-Deltic 18-cylinder diesel drives the pump. Also shown are the operator controls and gauges, and one of the two 12-inch suction hoses.

A left-side view of the Super Pumper engine and air intake.

The hydraulic crane to handle the large suction hose was replaced with a mechanical crane after delivery.

A front view of the F715FST tractor used on both the Super Pumper and Tender.

The Super Tender with the original McEntyre monitor which was later replaced with a Stang monitor.

The rear steering station of the Super Tender was removed after delivery. The monitor is shown in "action mode."

In this famous publicity shot, the Super Tender and all three satellites are discharging water through their large capacity monitors.

The three satellite hose tenders were C85 chassis with Mack diesels and manual transmissions. The 4,000-gpm monitors were fed by four 4 1/2-inch inlets. The hose bed could carry 2,000 feet of 4 1/2-inch hose.

The collector owned Super Pumper arrives at the 2004 SPAAMFAA national convention and muster in Frankenmuth, Michigan, pulled by a Mack Super-Liner tractor.

CHAPTER 11

The R Fire Series: 1966—1990

The R series commercial truck line was introduced in 1965, and like the B series it replaced, went on to become one of the world's most popular heavy-duty diesel trucks. When production of highway trucks ended in 1989, a total of 210,238 had been built, far eclipsing the totals of the venerable B models.

R model custom fire apparatus were first produced in 1966, but unlike their commercial brethren, did not outsell the predecessor B models. At this point in time, cab forward-type apparatus had captured the bulk of the market. American LaFrance had not offered a conventional-type custom since the mid-1940s, and Seagrave built their last conventional custom in 1970. The final custom R fire chassis came off the line in 1990, and was the last of 381 produced.

The R model fire apparatus featured a fiberglass tilt hood, a large two-piece curved windshield, and a bigger three-man cab. R model custom fire apparatus were identified by a serial number which consisted of an R for the model; a number 4 or 6 which, when two zeros were added, denoted the model series that was a relative indication of frame size; the next two numbers indicated the engine used; and the letter that followed indicated a pumper or fire chassis, and revealed if it was tandem equipped. The next number in parenthesis indicated the pump size to which two zeros must be added. Thus an R611FCS(10)1001 indicates an R model 600 series with an ENDT673 Thermodyne diesel, tandem axle fire chassis with a 1,000-gpm pump, and the first one built.

When first introduced, R model fire apparatus offered the following standard pumper specifications: ENF707C gasoline or ENDTF673C Thermodyne diesel power, a 5-speed transmission, air brakes,

This 1966 R608 was listed as a demonstrator for a Southern dealer.

manual steering, closed cab, 500-gallon tank, and a Waterous pump of 750-, 1000-, or 1,250-gpm capacities. Later, power steering became standard, and larger pumps, automatic, and synchromesh transmissions were available.

The gasoline engine was only available through 1973. In 1972, a small Mack-Scania diesel was available in a fire chassis. Only 23 of these were sold. As newer and higher horsepower Mack diesels became available, they too were offered. The biggest engine development was in 1966 when the Maxidyne engine was introduced. This engine proved very popular because of the substantially constant horsepower output available throughout the operating speed range. Benefits included reduced shifting and lower engine speeds at maximum pump capacities.

A subtle but significant change, and an indicator of things to come, marked the R model period. For the first time Mack was purchasing cabs from vendors and although some R models were built with Mack bodies, others were outsourced to other companies such as Hamerly and Ward 79. Mack advertising still touted that the engines and drive trains were Mack built but sheet metal parts and bodies were no longer listed as being built in Mack plants.

The bulk of R apparatus production was diesels with the standard coupe cab. In addition to pumpers, the R was widely used in tanker applications and some rescue rigs. Some tanker rigs were built on commercial chassis as well as on custom fire chassis. After 1980, the R was marketed as a custom fire chassis rather than a complete rig.

Two notable cab exceptions were involved with the R series. In 1969, five four-door cab 1,000-gpm pumpers were built for New York City. These were not popular because of poor maneuverability on city streets, especially compared with the CF models that became the standard styles for FDNY pumpers. These were the only R model pumpers used by FDNY, but eight R model rescues saw service with bodies by Theurer, Providence, Hamerly, and Pierce. The four-door sedan cab has made a comeback in recent years on commercial chassis by International and Freightliner, but Mack has not elected to compete in this market segment.

There was only one R model semi cab built, a 1967 R611F 750-gpm pumper, which was the sixth R611 built. Painted a dark blue, it is a very unique and attractive rig. Open cab apparatus have not been permitted since 1987 due to NFPA standards, and the last open cab custom pumper built is believed to be a 1982 American LaFrance pumper.

Throughout its 79-year history of custom fire apparatus production, Mack, unlike its major competitors, has always had a conventional-type rig in its lineup. The very last R model custom fire chassis built was an R688FC rescue truck with an E-One body delivered to Lawnton Fire Company in Pennsylvania on December 26, 1990, at a cost of $206,000.

R model custom fire apparatus designations were as follows:
All engines are diesel (except 608) and pump sizes ranged from 750 to 2,000 GPM.

Model	**Engine**	**# Built**
487	Scania ENDTF475 & ETZ477B 475ci 210 HP	23
608	Mack ENF707C Thermodyne 707ci 276 HP	61
611	Mack ENDTF673, E6-260, ETZ673 672ci 260 HP	75
612	Mack E6-315, ETAZ673 672ci 315 HP	2
685	Mack ENDT675, EM6237, ETZ675 Maxidyne 672ci 237 HP Mack EM6-250 672ci 250 HP	97
686	Mack ENDT676, EM6-285, ETSZ676 672ci 285 HP Mack EM6-300 672ci 300HP	59
688	Mack E6-350, ETAZ677 672ci 350 HP	49
690	Mack EM6-275 672ci 275 HP	15
		Total Built: 381

Only 61 R608s with the 707C Thermodyne gasoline engine were built. These engines developed 276 horsepower at 2,600 rpm.

The control panel for the 1,000-gpm Waterous pump shows an optional 2 1/2-inch gated inlet.

The standard hose body accommodated 1,200 feet of 2 1/2-inch hose and 500 feet of 1 1/2-inch hose.

This 1967 1,250-gpm R611 for Natchez, Louisiana, featured a Federal Q siren and individual discharge gauges.

The only R model sedan cabs built were five R611s for FDNY in 1969. They didn't compare favorably with the CF model which FDNY bought in large numbers from 1968 through 1989.

The R685 with the 237-horsepower Maxidyne engine was the best selling R model with 97 built. Rockville, Maryland, received four 1,000-gpm pumpers in 1969.

Bethel, Connecticut, received this R685 1,000-gpm pumper in 1971. Mack off-highway trucks can be seen in the background.

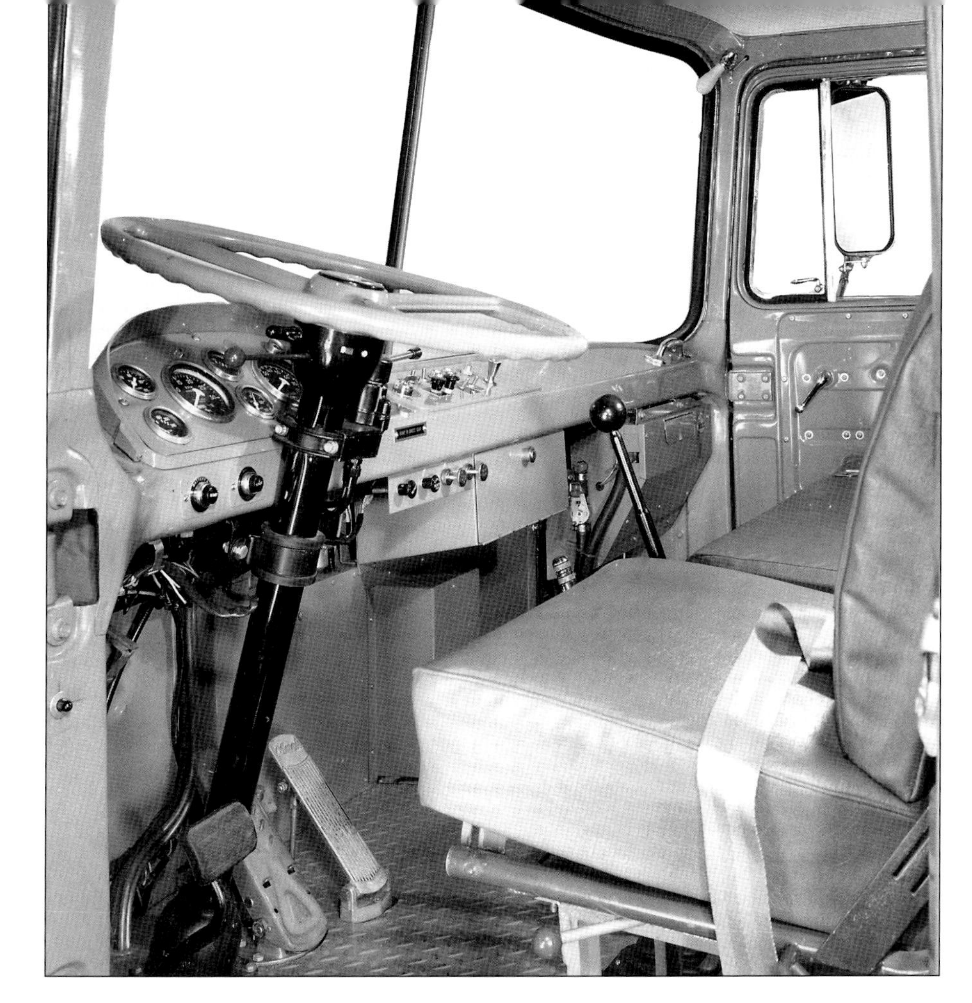

The cab area of this R model shows it was equipped with air brakes and a manual transmission.

The sixth R611 built in 1967 was the only R model equipped with an open semi cab. This dark blue 750-gpm pumper was delivered to Port Vue, Pennsylvania.

This 1972 R608 1,250-gpm pumper was originally delivered to LaFayette, Louisiana, but was in use at Galmey, Maryland, when this photo was taken.

Rincon Valley, California, received this R685 1,250-gpm pumper in 1973. *Bill Egan photo*

This 1975 commercial R model tractor pulls a 7,600-gallon tanker for the Fremont (New Hampshire) Fire Department.

Ranger built the body on this 1987 R chassis for Ashland, New Hampshire.

An R688FC was the chassis for this 1990 pumper by Emergency One. Milan, Vermont, specified a top-mount pump panel and numerous cross lays.

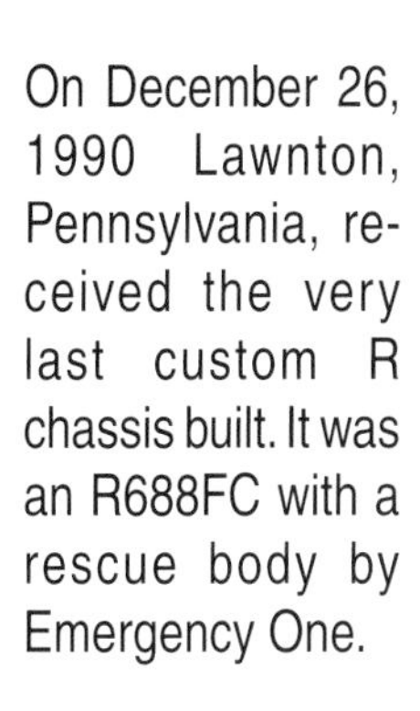
On December 26, 1990 Lawnton, Pennsylvania, received the very last custom R chassis built. It was an R688FC with a rescue body by Emergency One.

CHAPTER 12

The MB/MC Fire Series: 1972—1990

In 1963, the MB series of highway trucks was introduced for the city delivery market. Called both low cab-forward and cab-over designs, these trucks were ideal for a variety of light- and medium-duty tasks including refuse, airport refueling, local delivery, and light tractor work. Their exceptional maneuverability and high visibility features were well received by the trucking industry. When production ended in 1978, a total of 18,564 units had been produced.

The MC/MR series replaced the MB in 1978, and while not beauty contest winners, were more attractively styled than their predecessors. The steel and fiberglass cab tilted for easy engine servicing. Although they shared a common cab shell, the MC was designed for inner-city work either as a straight truck or tractor and its chassis design was based on the R model conventional truck. The MR was designed for construction and refuse applications, and featured extra heavy-duty components for arduous off-road service. The commercial MC was phased out in 1987 and the MR is still in production.

The MB custom fire units were not introduced until 1972 and 319 were built before production ended in 1978. The MC custom fire units were introduced in 1978 and the last of 756 units was built in 1990. The MR was never a custom fire chassis but many fire vehicles have been built on MR chassis up to and including the present.

This sales brochure announced the MB fire line, introduced in 1972.

Like the R series, the serial number designations of MB and MC custom fire apparatus indicated the engine used, pump size, and if a fire pumper, fire chassis, or fire tractor, in either four- or six-wheel configurations. A serial number of MC685FCS(10)1001 indicates the first MC 600 series tandem axle fire chassis with an ENDT675 Maxidyne diesel and a 1,000-gpm pump.

The MB series was available with two-man or five-man canopy cabs and two-stage centrifugal pump sizes from 750 to 1,250 gpm. Air brakes, power steering, and an alternator were standard. The MB was unique in two aspects. This was the only custom Mack apparatus to offer two non-Mack diesel engines; a 6-cylinder Scania, or a Caterpillar 3208 V-8. Only 57 Scanias and 4 Cats were produced. In addition, vendors, mainly Howe Fire Apparatus, built most MB bodies. This arrangement did not prove to be entirely satisfactory. A total of 319 MB models were built through 1978.

The MC custom fire apparatus was well received and 756 were built between 1978 and 1990. Only 28 were built with a Scania engine while five different Mack diesels were used. The MC was advertised as a first run, attack-type unit, and was well suited and received by those departments desiring apparatus of compact design.

Both Mack and other companies made MC bodies. Exotic canopies and fully enclosed crew cabs were added by a variety of fire apparatus builders. MCs were available with synchromesh, constant mesh, and automatic transmissions; but during the late 1980s, automatics were specified nearly 100 percent of the time. Pump sizes were available from 750 to 2,000 gpm.

MC models were popular as both pumpers and rescues, with some used as tankers and aerial tractors. The final MC built with a Mack body was a 1,500-gpm pumper delivered to Guttenberg, New Jersey, in 1984. As of December 2004 it was reported running as North Hudson Regional Squad 7.

The final MC custom fire chassis was the basis for a Saulsbury-bodied rescue delivered to Westbury, New York, on August 2, 1991, at a cost of $391,000. It came off the Mack assembly line on October 31, 1990, and was an MC688 FC model. An incredible coincidence is that this same fire district received the last CF pumper built with a Mack body in 1984.

The first MB685 was this canopy cab pumper built in 1973.

A breakdown of MB/MC custom fire models is as follows: The engines were all diesels and the pump sizes ranged from 750 to 2,000 GPM.

Model	Engine	# Built
MB487	Scania ENDT475, ETZ477B 475ci 210 HP	57
MC487	(same)	28
MB492	Caterpillar 3208 V-8 686ci 175 HP	4
MB611	Mack ENDT673, ETZ673, E6-260 Thermodyne 672ci 260 HP	79
MC611	(same)	183
MB685	Mack ENDT675, ETZ675, EM6-237, Maxidyne 672ci 237 HP	179
MC685	(same)	67
MC686	Mack ENDT676, ETSZ676, EM6-285 672ci 285 HP	316
MC688	Mack E6-350, ETAZ677 672ci 350 HP	142
MC690	Mack EM6-275 672ci 275 HP	20
		Total: 1075

The right side shot of the first MB685 shows optional front suction intake, Federal Q siren, and dual booster reels.

The MB driver's position featured large, flat windshields.

The tilt cab facilitated engine servicing.

This bare MB chassis shows the straight frame rails and 50-gallon rear-mounted fuel tank.

The MB body and pump panel were built and assembled for Mack by Howe Fire Apparatus.

Few MBs were built with the two-man cab, shown here in a two-tone paint job.

Fifty-seven MB487FCs were built with the small 475-cubic-inch Scania diesel. This one was delivered to North Middleton Twp. in Pennsylvania in 1974.

A few MBs were built with 50-foot Pitman Snorkels installed. This 1967 model ran as Snorkel Squad One in Chicago, Illinois.

Another California delivery was this 1974 MB685FC with Howe body for Alpine County. *Bill Egan photo*

The resourceful Chicago Fire Department shops constructed this giant deluge gun on an MB chassis. *William Stehman photo*

The MC custom fire models replaced the MB units in 1978. They were more attractively styled than the MB line. As noted in this head-on view, light bars and electronic sirens were now commonplace.

The pilot model MC685F sported a canopy cab and Mack body.

Aluminum ladders were now standard as were four enclosed side compartments and one in the rear.

Seat belts were provided for three firefighters riding in the canopy section.

The canopy section remained stationary with the front cab tilting 60 degrees, through the action of a single hydraulic cylinder.

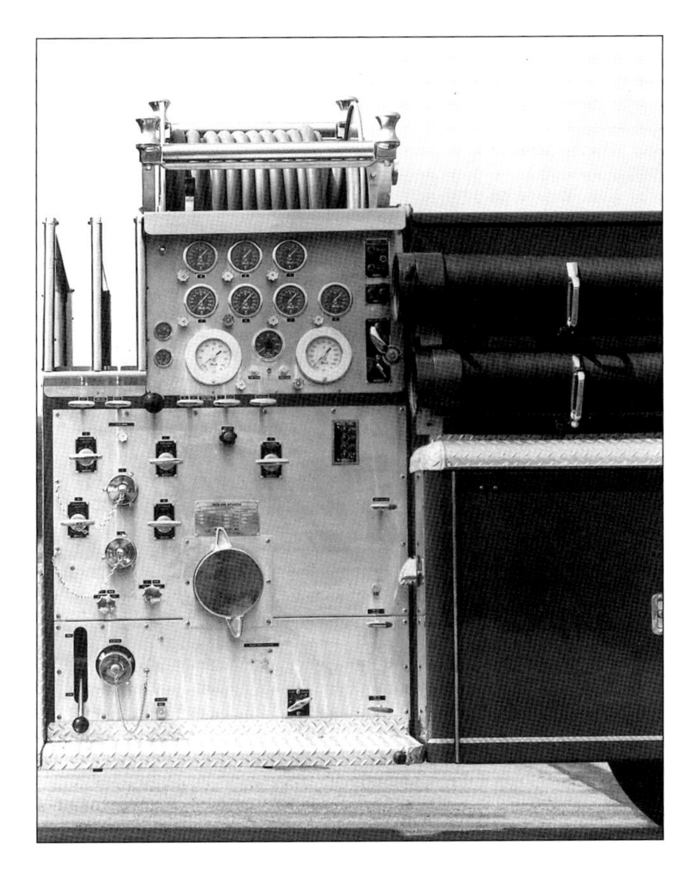

The final MC built with a Mack body was this 1,500-gpm pumper delivered to Guttenberg, New Jersey, in mid-1984. The serial number was MC686F(15)1155. It later became Engine 16 in the North Hudson Fire Department and at last report was listed as a North Hudson Regional reserve engine.

Featured on the MC685F pilot unit were individual discharge gauges, oversize master gauges, electric transfer valve, gated 2 1/2-inch inlet, cross lay compartment, and stainless steel pump panel.

Rescue Co. 5 of the FDNY responded in 1985 with this Mack MC-chassis rig with Saulsbury body. This was the first use of tandem rear axles on rescue rigs.

This 1,000-gpm MC685FC pumper with LTI body was delivered to Allentown, Pennsylvania, in 1986. *Dan Decher photo*

Four Guys built this 1,800-gallon pumper/tanker on an MC686FC chassis for Swannanoa, North Carolina, in 1987. *Dan Decher photo*

Camp Hill, Pennsylvania, was a Mack department for many years. On their roster was this 1983 MC with a Saulsbury rescue body.

The very last MC custom fire chassis built was this MC688FC, which came off the assembly line on October 31, 1990. Saulsbury installed a stainless steel rescue body, which was delivered to Westbury Fire District on Long Island, New York, on August 2, 1991.

Lebanon, Pennsylvania, area fire departments assembled these MC and MR pumpers and rescues for a group photo at a county fireman's convention in Avon, Pennsylvania, in 1994.

CHAPTER 13

The CF Fire Series: 1967—1990

The CF fire apparatus series was the second and final custom cab forward line built by Mack. It was Mack's "top of the line" fire series and became one of the most popular cab forward apparatus models ever. The CF was built from 1967 to 1984 as a complete fire apparatus, and from 1984 to 1990 as a custom fire chassis only. Over 3,800 units were produced in an incredible 24-year production run, making the CF the highest volume Mack fire model ever. CF production was begun at Allentown and moved to the new Macungie plant in 1976, where it remained until the end. An article in the Mack Bulldog magazine stated the Macungie plant, when opened, was able to complete two fire trucks per day, and that Mack was the largest producer of fire apparatus in the country.

The CF represented Mack very well in the highly competitive custom cab forward market. It was sold in all areas of the United States. It was used by one-unit volunteer departments up to and including the largest fire department in the country, New York City. Pumpers were the main units produced, but the line included straight frame aerials, in both mid-mount and rear-mount configurations, as well as tractor-drawn tillers. Hundreds of aerial platforms were produced under the trade name Aerialscope. The CF was also used as the chassis for heavy-duty rescue and special purpose units.

The CF cab was an ingenious adaptation of the over-the-road F model cab produced from 1962 through 1981. The forward section and canopy were built by Mack initially, but later would be purchased

The very first CF built was this pilot car CF608F(10)1000 in 1967. It had a 707C gasoline engine, and a 1,000-gpm Waterous pump. It was a demonstrator before being sold to Moraine, Ohio, in 1968. It was then sold to Jefferson Twp. Fire Department near Dayton where considerable changes were made to the body. As of this writing it was still on the active roster and retains its original engine.

from a vendor. Fire departments widely praised the CF cab for its high visibility feature and its roominess.

During the reign of the CF, Mack also offered the R model conventional-type apparatus; the MB/MC/MR low cab-over engine design; and the Mid-Liner C.O.E. and conventional series. The CF was the apparatus of choice, however, when powerful performance, spacious cabs, and good maneuverability were required.

Air brakes and power steering were always standard on the CF. The 707C gasoline motor was available until 1973. Mack diesels of various horsepower and size were the principal power sources. Waterous pumps of various capacities and stages were standard. Mack transmissions were standard but vendors supplied synchromesh and automatic transmissions. About a dozen open cab CF models were produced, but virtually the total production consisted of closed cabs with open or closed canopy sections.

The key to decoding serial numbers was as follows: CF denoted the basic model, cab forward; the first digit denoted the model series 6 or 7, an indication of frame size, to which two zeros were added indicating 600 or 700; the following two digits designated the engine model; the suffix letter designated the type of apparatus, F for fire pumper, FAP for fire aerial platform, FAPS for aerial platform with tandem axles, and FCA or FA for aerial ladder. The suffix numbers designated pump size, to which two zeros were added to indicate the actual pump capacity; and the remaining numbers indicated the sequential serial numbers of each unit. As an example, CF611F(10)1001 denoted a CF600 pumper with an E6-260 Thermodyne diesel, 1,000-gpm pump, and serial number 1001, the first unit built.

This Maxidyne powered CF685F was built in 1970 for a dealer in Idaho. The front end stainless steel sheet greatly improved on the original painted front end with horizontal chrome strips.

The highly successful Mack aerial platform with the trade name of Aerialscope (called a tower ladder by most departments), which was first introduced in the C series in 1964, continued to be a prominent part of the CF production. The first boom built by Baker Equipment Engineering Co. of Richmond, Virginia, was mounted on a CF chassis in 1970 for delivery to Bedford, Ohio. Between 1970 and 1987 Baker would build 340 75-foot scopes on Mack chassis (155 for FDNY and 185 for others) and 35 95-foot scopes on Mack chassis (17 for FDNY and 18 for others), for a total of 375 units. After 1982 all FDNY scopes had tandem axles. The first 95-foot unit was delivered to Providence, Rhode Island, in 1985.

CF models available were as follows: All engines are Mack diesels (except 608), and pump sizes ranged from 750 to 2,000 GPM.

Model	**Engine**	**# Built**
608	ENF707C Thermodyne gasoline 707ci 276 HP	173
611	ENDT673C, E6-260, ETZ673 Thermodyne 672ci 260 HP	1270
612	E6-315, ETAZ673 672ci 315 HP	24
685	ENDT675, EM6-237, ETZ675 Maxidyne 672ci 237 HP	1216
686	ENDT676, EM6-285, ETSZ676 672ci 285 HP	625
688	E6-350, ETAZ677 672ci 350 HP	426
690	EM6-275 672ci 275 HP	3
719	ENDTF864 V-8 864ci 300 HP	14
795	ENDTF865 V-8 865ci 325 HP	86
797	ENDT866 V-8 866ci 375 HP	12
		Total Built: 3,849

Small 750-gpm pumps were rare on CFs but this 1970 CF685F delivery for a western Pennsylvania department was so equipped. The customary front-mounted siren is missing.

In 1988, Mack gave Baker the exclusive rights to market the Aerialscope through the Mack network or direct to customers. After the demise of the Mack CF chassis, Baker considered using chassis by FWD/Seagrave, Spartan, Simon-Duplex, and Autocar.

The first Baker scope mounted on a non-Mack chassis was on a Simon-Duplex in 1991.

Baker filed for bankruptcy protection in 1995, and in 1997 sold the Aerialscope division to a company associated with FWD/Seagrave; in 2003 the operation was moved from Virginia to the Seagrave factory in Clintonville, Wisconsin. In a humiliating turn of events the Mack-pioneered aerial platform is now controlled and marketed by a former archrival.

It is entirely fitting that the last Aerialscope produced by Mack was delivered to FDNY on December 30, 1991; thus ending a reign of 28 years of the Mack Aerialscope in FDNY. It was a CF688FCS model that cost $504,248.

Pumpers were the principal CF product, but aerials made up a healthy share of Mack orders. Maxim was the primary supplier of aerial ladders to Mack since 1949, but many customers opted for Pirsch, Snorkel, and LTI units as well. In the early 1970s Thibault of Canada became the standard aerial ladder supplier; but like Maxim, not on an exclusive basis. Parts and service problems plagued this alliance and in the late 1970s the unthinkable happened. Mack began to buy ladders from long-time rival American LaFrance. This "unholy alliance" also didn't last long, and in 1981 Mack introduced the Bulldog I 106-foot ladder made exclusively for Mack by an LTI company, Conestoga Custom Products, Inc. These 4-section all-steel ladders were available in mid-rear or tractor-drawn configurations, and were used until Mack exited the custom fire business. The first Bulldog I delivery was to Ogunquit, Maine, and the first tiller was delivered to Downey, California.

There was considerable turmoil in the fire apparatus industry during the latter half of the twentieth century with several old-line builders going out of business. The "handwriting was on the wall" at Mack when, in 1980, an emphasis was placed on the sale of fire chassis only rather than complete rigs.

In 1984, Mack announced they would no longer build apparatus bodies in their own plants, which had been done for the prior 70 years at factories in Allentown and Macungie, Pennsylvania, Long Island City, New York, and Sidney, Ohio. The final Mack-bodied pumper was a CF688F delivered to Westbury, New York, on August 21, 1984. It had a 1,500-gpm pump and a 600-gallon tank, and cost $179,855.

The CF, MC, and R model custom fire chassis continued to be built at the Macungie plant until 1990. On January 5, 1990, Mack notified its distributors and branch managers that it would discontinue the production of R, MC, and CF fire apparatus chassis. The "official" reason given was that the minimum volumes did not support the continued engineering and future development investment and that the requirements for engine certification for the 1991 calendar year could not be achieved with the present engine — the cost to engineer the E7 engine into this chassis would be cost prohibitive. Thus, with a simple, terse announcement Mack withdrew from a market they had competed in with considerable prominence for nearly 80 years.

Additional factors, which contributed to the decision to cease fire chassis production were: (1) negative financial pressure, including labor costs, were affecting Mack as well as others, (2) customers were demanding a choice of bodies and cabs made of steel, aluminum, or stainless steel and, (3) apparatus was becoming far more complicated with the advent of new foam systems and complex electronic controls.

After the cessation of body building by Mack, mongrelization of fire apparatus reached new heights. Almost every one of Mack's former competitors completed units on Mack chassis, utilizing every make of pump available. Besides bodies, some manufacturers made exotic modifications to the Mack cab. Saulsbury was especially adept at creating "hi rise command" cabs on the standard Mack CF structure.

Mack advertising literature prior to 1984 encouraged the buyer to buy an apparatus completely built in the plant of the manufacturer and to avoid the pitfall of "hodge-podge" assembly. With the discontinuance of building their own bodies after 1984, Mack disregarded their own advice and the construction of the last CF pumper was a painful reminder of the pitfalls of divided responsibilities. The final CF pumper chassis came off the Mack assembly line in December 1990 but the complete unit was not delivered until an agonizing 20 months later. Originally the body was to be built by Ward 79 of Elmira Heights, New York, which went out of business prior to delivery of the chassis. The body was finally built and installed by Interstate Truck Equipment Inc. of Hagerstown, Maryland. Although the final result was quite satisfactory, the time involved was considerable.

The Yonkers Fire Department placed the last CF pumper in service on August 8, 1992, a scant two days after delivery. It was a CF688FC 1,000-gpm pumper with a 350-hp engine and an Allison HT740 automatic transmission. The cost was $184,501.

The demise of the CF was a particularly bitter pill for the Mack enthusiast to swallow since the CF was an exclusive fire service vehicle and represented Mack very well in this demanding, high profile market. Its distinctive appearance and motor sound made it very recognizable on the fire scene. During recent years, Mack was the only custom builder to use its own engines and drive train components.

A replacement for the CF was actually proposed and drawings were made, but it never even made it to the prototype stage. Mack fans have been hoping that Mack would re-enter the apparatus field but it appears fire apparatus has joined buses and off-highway trucks as permanent deletions from the Mack product line.

A CF611F chassis moved down the Allentown plant assembly line in 1968.

This 1970 CF611FAP was the first Aerialscope on a CF chassis and the first boom and platform built by Baker Equipment Engineering Co. The Mack World Headquarters Building in Allentown officially opened in 1970 also.

This is believed to be the only CF open cab Mack/Maxim aerial ladder truck built. It was a 1970 with a 100-foot Maxim mid-mount ladder for Riverdale, Maryland.

Only three of these exotic airport crash truck rigs were built. This 1971 CF685F went to Glenmont, New York.

The roomy CF cab was a clever adaptation of the over-the-road F model cab. The F model was produced from 1962 through 1981.

In 1971 FDNY received 40 CF611F(10) pumpers. This was the first order with an enclosed canopy section. In 1972 every first line pumper in the department was a Mack.

This CF685F was Mack's first and last venture into stainless steel body construction. It was sold to Blue Ash, Ohio, and in 2005 was for sale by an Ohio apparatus dealer.

Mack built 112 CFs with V-8 diesel engines. This engine makes for a full compartment. Chrome valve covers were a long time Mack fire apparatus engine feature.

This 1974 CF685F for Richland, Pennsylvania, sports a new two-tone paint combination which became a very popular option.

One of the last factory photos in the files was this CF685F(12) for Chesterfield Co., Virginia, in 1979. A 1,250-gpm pump was furnished as well as twin roof beacons.

The popular Mack/Maxim aerial ladder line, as shown here, was phased out in the early seventies.

Mack teamed up with archrival American LaFrance for this 1978 CF685FCA chassis with a rear-mount LaFrance ladder for delivery to Barren Hill, Pennsylvania.

This is believed to be the only Mack/LaFrance tiller built. The tractor was a CF Mack, the ladder was an American LaFrance and the bodywork was by Hamerly.

Mack was a popular chassis for Pitman Snorkels. This 1970 unit had a Mack CF chassis, a Pitman Snorkel, and bodywork by Pirsch. Milwaukee and Glendale, Wisconsin, were the operators of this rig.

Milwaukee, Wisconsin, liked Mack chassis and Pirsch ladders. This 1977 CF with a 100-foot Pirsch ladder was a popular rig.

The final ladder manufacturer for Mack was LTI. This 1982 CF686FA chassis was the basis for a 106-foot LTI ladder with bodywork by Conestoga. *Dan Decher photo*

Mack built less than a dozen open cab CF models. This CF611F(12) was delivered to Cumberland, Maryland, in 1968.

This is believed to be the last open cab CF built in 1974. It was a CF685F(10), the last of three open cab CFs delivered to Connellsville, Pennsylvania. It was sold to Dunbar, Pennsylvania, and is presently owned by a collector.

Lynn, Massachusetts, operated this 1970 CF611FT pulling a 100-foot Maxim aerial ladder.

This 1972 CF685F(12) saw service in Pierce County, Washington. *Glenn Vincent photo*

The last of 173 CFs with the 707C Thermodyne gasoline engine was this CF608F(10) delivered to Earlville, New York, in 1973. It is now owned by a collector who has plans for a museum in the future.

Replacing a 1951 Mack 505A pumper (which replaced a 1925 Mack AC pumper) was this 1978 CF611(10) white pumper with red wheels, in service with the Defender Fire Co. in Berwick, Pennsylvania.

It was a sad day for Mack fire apparatus enthusiasts when the last pumper with a Mack body rolled off the assembly line in 1984. It was a 1,500-gpm pumper with serial number CF688F(15)1067 and cost $179,855.00. It has been in service in Westbury, New York, but is due to be replaced in 2005. *Glenn Usdin photo*

When Mack no longer built a complete truck, many departments that favored Mack had another company build on the Mack chassis. Mastersonville, Pennsylvania, had Pierce build a massive pumper/tanker on this Mack CF688FCS chassis. *Mastersonville Fire Department photo*

On December 12, 1990 Mack completed its last custom CF Aerialscope chassis. It was a CF688FCS model and on December 30, 1991 FDNY accepted delivery. On March 3, 1992 it went in service as Ladder 58, and was the last of 172 Aerialscopes delivered to FDNY on Mack chassis since 1964. *Joe Pinto/Fire Apparatus Journal photo*

The horn button on the last Aerialscope delivered reads: "Custom Built by Mack for City of New York Fire Department." This was normal procedure when the customer was known at the time of manufacture.

The very last custom CF pumper chassis built was a CF688FC which came off the line on December 18, 1990. The body was to be built by Ward 79, who was no longer in business. The 1,000-gpm pumper was finished by Interstate Truck Equipment Inc. of Hagerstown, Maryland, and delivered to Yonkers, New York, on August 6, 1992. As of February 2005 it was in reserve status and plans call for it to be part of a fire department museum in the future. *Tom Adams photo*

CHAPTER 14

Miscellaneous

In addition to the major model fire apparatus series' detailed in the previous chapters, just about every Mack model saw some application as a fire vehicle. Some were only used on a very special basis and in very limited numbers.

In 1936 and 1937 Mack enlisted the Reo Motor Car Company to build 4,974 light and medium trucks for sale through some Mack dealers. These were built in both conventional and cab-over styles. These trucks were named the Mack Jr. line. A small number were built as fire trucks but did not carry custom fire apparatus serial numbers.

An N model was introduced in 1958 to replace the D series. These were of the low cab-over configuration and used the same Budd built cab as used on the Ford C series trucks. A total of 1,945 trucks were produced through 1962. Five were produced between 1960 and 1963 as custom fire apparatus and carried custom fire apparatus model designations N85F, N95F, and N505F.

In 1981, a Renault RVI line of medium-duty pumpers were made available, branded as Mack Mid-Liner MS models MS200P and MS300P. A seven-man cab and canopy module were standard equipment. These were sold as an option between commercial and custom built apparatus. These were initially built exclusively for Mack by Ward 79 and featured an aluminum body. The first delivery was to Tamaqua, Pennsylvania. In 1984, the Mid-Liner conventional-type CS series became available as fire vehicles. A total of 285 Mid-Liners were built, but Mack apparatus fans never regarded these models as "real" Macks.

While it is true that Mack no longer builds a complete fire apparatus or offers a custom fire chassis, this does not mean the end of Mack in this market. Although the following alternatives do not enthrall

The exact number of fire trucks on the Mack Jr. chassis is not known, but this was a 1936 open cab pumper. Reo made this model for Mack in 1936 and 1937.

the devoted Mack enthusiast, they will, nonetheless, maintain a Mack presence:

1) All Mack chassis are available as "commercial" chassis for another manufacturer to build as a complete fire vehicle. Although this option involves more extensive modifications by the final assembler, a considerable number of R construction vehicle chassis, MR chassis, CH conventional chassis, and the recent Granite series chassis are being adapted to fire service use.

2) For the first time in their history, Mack is making their engines and chassis components available to other original equipment manufacturers. One of the first Mack diesel engines sold to another manufacturer for use in their chassis was installed in a Simon-Duplex chassis in 1991. It was an E7 300-hp engine mated to an Allison HT-740 automatic transmission. Simon-Duplex is no longer in business.

This unit was identified as a Type 309 on a Mack Jr. 30MB chassis. Someone went to a lot of effort to cast the Mack Jr. logo on both the intake and discharge caps.

Five N model custom fire apparatus were built between 1960 and 1963. This N505F was a floodlight wagon for Hicksville (New York) Fire Department.

This N85F pumper was assembled by Hahn for Mack in 1960. Pump controls indicate a 750-gpm Waterous pump.

This rear shot of the Marcus Hook rig shows a full complement of aluminum ladders.

The last of three N85Fs was this 1963 open cab aerial for Marcus Hook, Pennsylvania.

The first Mid-Liner pumper was delivered to Tamaqua, Pennsylvania, in 1981. Ward 79 built the bodies and this series never really caught on.

Canopy cabs were more popular than this three-man, two-door, tilt-type closed cab Mid-Liner.

Los Angeles, California, ran this 1975 heavy rescue based on a commercial FL series, first produced in 1966. *Bill Egan photo*

Not all fire trucks are loaded with chrome and gold leaf. This 1975 brush and tank wagon was a former military rig and was locally built by the Mount Carbon (Pennsylvania) Fire Department. *Dan Decher photo*

MH commercial models were built from 1983 to 1990, and this 1987 rescue with a Swab body was the only one I know of that was used as a fire vehicle. *Dan Decher photo*

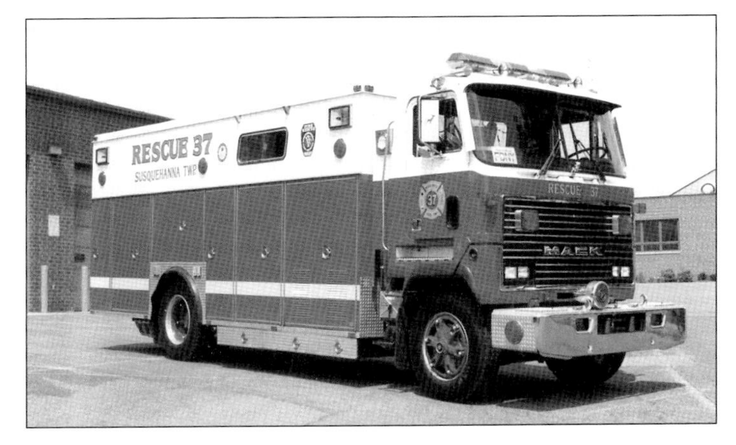

This is the only photo in this book that is not on a Mack chassis. It does, however, have a Mack E-7 300-horsepower engine, one of the first sold to another manufacturer. New Lexington mounted the body on this Simon-Duplex chassis in 1992.

The tilt cab of this 1992 Simon-Duplex shows the Mack engine installed in a competitor chassis.

St. Marys, Pennsylvania, had the Snorkel boom from a 1966 C model Mack transferred to this MR commercial chassis in 1992. New Lexington did the assembly work. *Dan Decher photo*

Tamaqua, Pennsylvania, liked their previous Mid-Liner well enough to have KME build this 1,250-gpm pumper on a Mid-Liner chassis in 1996. *Dan Decher photo*

Morrilton, Arkansas, had KME build this 1998 1,250-gpm pumper with top mount controls on a commercial MR688P chassis. *Dan Decher photo*

This super sized 1,500-gpm, 2,600-gallon pumper/tanker was built by Swab in 2000 on a MR688P chassis for Harleysville, Pennsylvania. *Dan Decher photo*

The 1993 Satellite 1 of FDNY was destroyed in the World Trade Center attack. Saulsbury built this replacement on a donated Mack MR chassis in 2002. *John A. Calderone photo*

In 2002 KME built this 1,500-gpm, 3,000-gallon pumper/tanker on the recently introduced Granite CV713 chassis for N. Cornwall Twp., Pennsylvania. *Dan Decher photo*

The RD series construction line is the basis for many fire vehicles. This 2003 RD688P was the chassis for this KME delivery to Blain, Pennsylvania. *Dan Decher photo*

Four Guys Fire Trucks built this classy tanker in 2004 for the Canton (Pennsylvania) Fire Department. It has a 1,250-gpm pump, a 4,000-gallon tank and lots of chrome. The chassis is a Granite CV713 which is reported to be the best selling heavy-duty class 8 conventional straight truck in the U.S. *Four Guys photo*

The latest addition to the roster of the Keystone H & L Co. of Myerstown, Pennsylvania, is this 2005 Mack MR688P chassis with a New Lexington rescue body.

More great titles from **Iconografix**

EMERGENCY VEHICLES

100 Years of American LaFrance: An Illustrated History ISBN 1-58388-139-5
The American Ambulance 1900-2002: An Illustrated History ISBN 1-58388-081-X
American Fire Apparatus Co. 1922-1993 Photo Archive ISBN 1-58388-131-X
American Funeral Vehicles 1883-2003 Illustrated History ISBN 1-58388-104-2
American LaFrance 700 Series 1945-1952 Photo Archive ISBN 1-882256-90-5
American LaFrance 700 Series 1945-1952 Photo Archive Volume 2 ISBN 1-58388-025-9
American LaFrance 700 & 800 Series 1953-1958 Photo Archive ISBN 1-882256-91-3
American LaFrance 900 Series 1958-1964 Photo Archive ISBN 1-58388-002-X
Classic Seagrave 1935-1951 Photo Archive ISBN 1-58388-034-8
Crown Firecoach 1951-1985 Photo Archive ISBN 1-58388-047-X
Encyclopedia of Canadian Fire Apparatus ISBN 1-58388-119-0
Fire Chief Cars 1900-1997 Photo Album ISBN 1-882256-87-5
Firefighting Tanker Trucks and Tenders: A Fire Apparatus Photo Gallery ISBN 1-58388-138-7
FWD Fire Trucks 1914-1963 Photo Archive ISBN 1-58388-156-5
Hahn Fire Apparatus 1923-1990 Photo Archive ISBN 1-58388-077-1
Heavy Rescue Trucks 1931-2000 Photo Gallery ISBN 1-58388-045-3
Imperial Fire Apparatus 1969-1976 Photo Archive ISBN 1-58388-091-7
Industrial and Private Fire Apparatus 1925-2001 Photo Archive ISBN 1-58388-049-6
*Mack Model L Fire Trucks 1940-1954 Photo Archive** ISBN 1-882256-86-7
Mack Fire Trucks 1911-2005 Illustrated History ISBN 1-58388-157-3
Maxim Fire Apparatus 1914-1989 Photo Archive ISBN 1-58388-050-X
Maxim Fire Apparatus Photo History ISBN 1-58388-111-5
Navy & Marine Corps Fire Apparatus 1836 -2000 Photo Gallery ISBN 1-58388-031-3
Pierre Thibault Ltd. Fire Apparatus 1918-1990 Photo Archive ISBN 1-58388-074-7
Pirsch Fire Apparatus 1890-1991 Photo Archive ISBN 1-58388-082-8
Police Cars: Restoring, Collecting & Showing America's Finest Sedans ISBN 1-58388-046-1
Saulsbury Fire Rescue Apparatus 1956-2003 Photo Archive ISBN 1-58388-106-9
Seagrave 70th Anniversary Series Photo Archive ISBN 1-58388-001-1
Seagrave Fire Apparatus 1959-2004 Photo Archive ISBN 1-58388-132-8
TASC Fire Apparatus 1946-1985 Photo Archive ISBN 1-58388-065-8
Van Pelt Fire Apparatus 1925-1987 ISBN 1-58388-143-3
Volunteer & Rural Fire Apparatus Photo Gallery ISBN 1-58388-005-4
W.S. Darley & Co. Fire Apparatus 1908-2000 Photo Archive ISBN 1-58388-061-5
Wildland Fire Apparatus 1940-2001 Photo Gallery ISBN 1-58388-056-9
Young Fire Equipment 1932-1991 Photo Archive ISBN 1-58388-015-1

BUSES

Buses of ACF Photo Archive Including ACF-Brill And CCF-Brill ISBN 1-58388-101-8
Buses of Motor Coach Industries 1932-2000 Photo Archive ISBN 1-58388-039-9
City Transit Buses of the 20th Century Photo Gallery ISBN 1-58388-146-8
Fageol & Twin Coach Buses 1922-1956 Photo Archive ISBN 1-58388-075-5
Flxible Intercity Buses 1924-1970 Photo Archive ISBN 1-58388-108-5
Flxible Transit Buses 1953-1995 Photo Archive ISBN 1-58388-053-4
GM Intercity Coaches 1944-1980 Photo Archive ISBN 1-58388-099-2
Greyhound in Postcards: Buses, Depots and Posthouses ISBN 1-58388-130-1
Highway Buses of the 20th Century Photo Gallery ISBN 1-58388-121-2
*Mack® Buses 1900-1960 Photo Archive** ISBN 1-58388-020-8
New York City Transit Buses 1945-1975 Photo Archive ISBN 1-58388-149-2
Prevost Buses 1924-2002 Photo Archive ISBN 1-58388-083-6
Trailways Buses 1936-2001 Photo Archive ISBN 1-58388-029-1
Trolley Buses 1913-2001 Photo Archive ISBN 1-58388-057-7
Yellow Coach Buses 1923-1943 Photo Archive ISBN 1-58388-054-2

TRUCKS

AM General: Hummers, Mutts, Buses & Postal Jeeps ISBN 1-58388-135-2
Autocar Trucks 1899-1950 Photo Archive ISBN 1-58388-115-8
Autocar Trucks 1950-1987 Photo Archive ISBN 1-58388-072-0
Beverage Trucks 1910-1975 Photo Archive ISBN 1-882256-60-3
*Brockway Trucks 1948-1961 Photo Archive** ISBN 1-882256-55-7
Chevrolet El Camino Photo History Incl. GMC Sprint & Caballero ISBN 1-58388-044-5
Circus and Carnival Trucks 1923-2000 Photo Archive ISBN 1-58388-048-8
Dodge B-Series Trucks Restorer's & Collector's Reference Guide and History ISBN 1-58388-087-9
Dodge C-Series Trucks Restorer's & Collector's Reference Guide and History ISBN 1-58388-140-9
Dodge Pickups 1939-1978 Photo Album ISBN 1-882256-82-4
Dodge Power Wagons 1940-1980 Photo Archive ISBN 1-882256-89-1
Dodge Power Wagon Photo History ISBN 1-58388-019-4
Dodge Ram Trucks 1994-2001 Photo History ISBN 1-58388-051-8
Dodge Trucks 1929-1947 Photo Archive ISBN 1-882256-36-0
Dodge Trucks 1948-1960 Photo Archive ISBN 1-882256-37-9
Ford 4x4s 1935-1990 Photo History ISBN 1-58388-079-8
Ford Heavy-Duty Trucks 1948-1998 Photo History ISBN 1-58388-043-7
Ford Ranchero 1957-1979 Photo History ISBN 1-58388-126-3
Freightliner Trucks 1937-1981 Photo Archive ISBN 1-58388-090-9
FWD Trucks 1910-1974 Photo Archive ISBN 1-58388-142-5
GMC Heavy-Duty Trucks 1927-1987 ISBN 1-58388-125-5
Jeep 1941-2000 Photo Archive ISBN 1-58388-021-6
Jeep Prototypes & Concept Vehicles Photo Archive ISBN 1-58388-033-X
Kenworth Trucks 1950-1979 At Work ISBN 1-58388-147-6
*Mack Model AB Photo Archive** ISBN 1-882256-18-2
*Mack AP Super-Duty Trucks 1926-1938 Photo Archive** ISBN 1-882256-54-9
*Mack Model B 1953-1966 Volume 2 Photo Archive** ISBN 1-882256-34-4
*Mack EB-EC-ED-EE-EF-EG-DE 1936-1951 Photo Archive** ISBN 1-882256-29-8
*Mack FC-FCSW-NW 1936-1947 Photo Archive** ISBN 1-882256-28-X
*Mack FG-FH-FJ-FK-FN-FP-FT-FW 1937-1950 Photo Archive** ISBN 1-882256-35-2
*Mack LF-LH-LJ-LM-LT 1940-1956 Photo Archive** ISBN 1-882256-38-7
*Mack Trucks Photo Gallery** ISBN 1-882256-88-3
New Car Carriers 1910-1998 Photo Album ISBN 1-882256-98-0
Peterbilt Trucks 1939-1979 At Work ISBN 1-58388-152-2
Refuse Trucks Photo Archive ISBN 1-58388-042-9
Studebaker Trucks 1927-1940 Photo Archive ISBN 1-882256-40-9
White Trucks 1900-1937 Photo Archive ISBN 1-882256-80-8

RECREATIONAL VEHICLES & OTHER

Commercial Ships on the Great Lakes Photo Gallery ISBN 1-58388-153-0
Phillips 66 1945-1954 Photo Archive ISBN 1-882256-42-5
RVs & Campers 1900-2000: An Illustrated History ISBN 1-58388-064-X
Ski-Doo Racing Sleds 1960-2003 Photo Archive ISBN 1-58388-105-0
The Collector's Guide to Ski-Doo Snowmobiles ISBN 1-58388-133-6

AUTOMOTIVE

AMC Cars 1954-1987: An Illustrated History ISBN 1-58388-112-3
AMC Performance Cars 1951-1983 Photo Archive ISBN 1-58388-127-1
AMX Photo Archive: From Concept to Reality ISBN 1-58388-062-3
Auburn Automobiles 1900-1936 Photo Archive ISBN 1-58388-093-3
Camaro 1967-2000 Photo Archive ISBN 1-58388-032-1
Checker Cab Co. Photo History ISBN 1-58388-100-X
Chevrolet Corvair Photo History ISBN 1-58388-118-2
Chevrolet Station Wagons 1946-1966 Photo Archive ISBN 1-58388-069-0
Classic American Limousines 1955-2000 Photo Archive ISBN 1-58388-041-0
The Complete U.S. Automotive Sales Literature Checklist 1946-2000 ISBN 1-58388-155-7
Cord Automobiles L-29 & 810/812 Photo Archive ISBN 1-58388-102-6
Corvair by Chevrolet Experimental & Production Cars 1957-1969, Ludvigsen Library Series ISBN 1-58388-058-5
Corvette The Exotic Experimental Cars, Ludvigsen Library Series ISBN 1-58388-017-8
Corvette Prototypes & Show Cars Photo Album ISBN 1-882256-77-8
Duesenberg Racecars and Passenger Cars Photo Archive ISBN 1-58388-145-X
Early Ford V-8s 1932-1942 Photo Album ISBN 1-882256-97-2
Ferrari- The Factory Maranello's Secrets 1950-1975, Ludvigsen Library Series ISBN 1-58388-085-2
Ford Postwar Flatheads 1946-1953 Photo Archive ISBN 1-58388-080-1
Ford Station Wagons 1929-1991 Photo History ISBN 1-58388-103-4
Hudson Automobiles 1934-1957 Photo Archive ISBN 1-58388-110-7
Imperial 1955-1963 Photo Archive ISBN 1-882256-22-0
Imperial 1964-1968 Photo Archive ISBN 1-882256-23-9
Jaguar XK120, XK140, XK150 Sports Cars Ludvigsen Library Series ISBN 1-58388-150-6
Javelin Photo Archive: From Concept to Reality ISBN 1-58388-071-2
The Lincoln Continental Story: From Zephyr to Mark II ISBN 1-58388-154-9
Lincoln Motor Cars 1920-1942 Photo Archive ISBN 1-882256-57-3
Lincoln Motor Cars 1946-1960 Photo Archive ISBN 1-882256-58-1
Mercedes-Benz 300SL: Gullings and Roadsters 1954-1964 Ludvigsen Library Series ISBN 1-58388-137-9
MG Saloons and Coupes 1925-1980 ISBN 1-58388-144-1
Nash 1936-1957 Photo Archive ISBN 1-58388-086-0
Packard Motor Cars 1946-1958 Photo Archive ISBN 1-882256-45-X
Pontiac Dream Cars, Show Cars & Prototypes 1928-1998 Photo Album ISBN 1-882256-93-X
Pontiac Firebird Trans-Am 1969-1999 Photo Album ISBN 1-882256-95-6
Pontiac Firebird 1967-2000 Photo History ISBN 1-58388-028-3
Rambler 1950-1969 Photo Archive ISBN 1-58388-078-X
Stretch Limousines 1928-2001 Photo Archive ISBN 1-58388-070-4
Studebaker 1933-1942 Photo Archive ISBN 1-882256-24-7
Studebaker Hawk 1956-1964 Photo Archive ISBN 1-58388-094-1
Studebaker Lark 1959-1966 Photo Archive ISBN 1-58388-107-7
Ultimate Corvette Trivia Challenge ISBN 1-58388-035-6

All Iconografix books are available from direct mail specialty book dealers and bookstores worldwide, or can be ordered from the publisher. For book trade and distribution information or to add your name to our mailing list and receive a **FREE CATALOG** contact:

Iconografix, Inc.
PO Box 446, Dept BK
Hudson, WI, 54016

Telephone: (715) 381-9755,
(800) 289-3504 (USA),
Fax: (715) 381-9756
info@iconografixinc.com

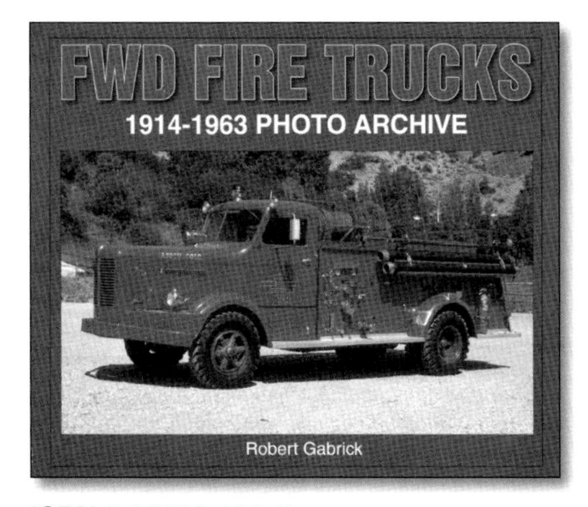

ISBN 1-58388-156-5

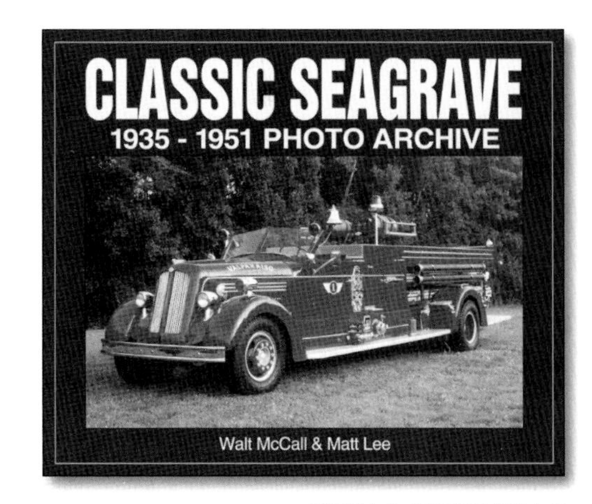

ISBN 1-58388-034-8

More great books from
Iconografix

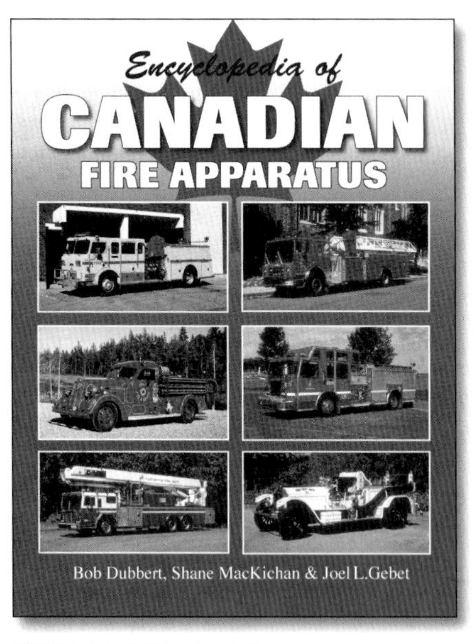

ISBN 1-58388-119-0

Iconografix, Inc.
P.O. Box 446, Dept BK,
Hudson, WI 54016
For a free catalog call:
1-800-289-3504
info@iconografixinc.com

ISBN 1-58388-139-5

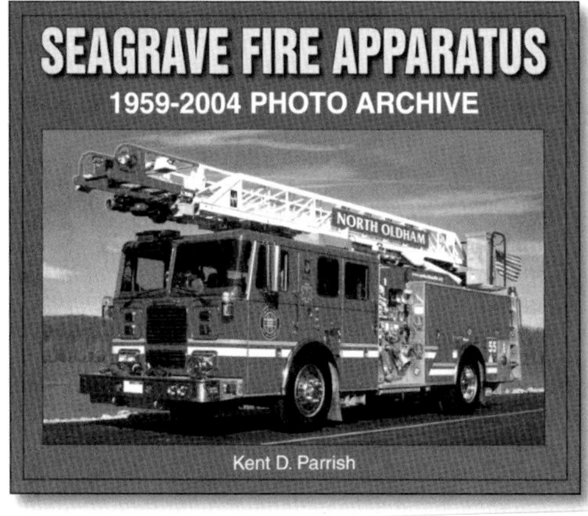

ISBN 1-58388-132-8

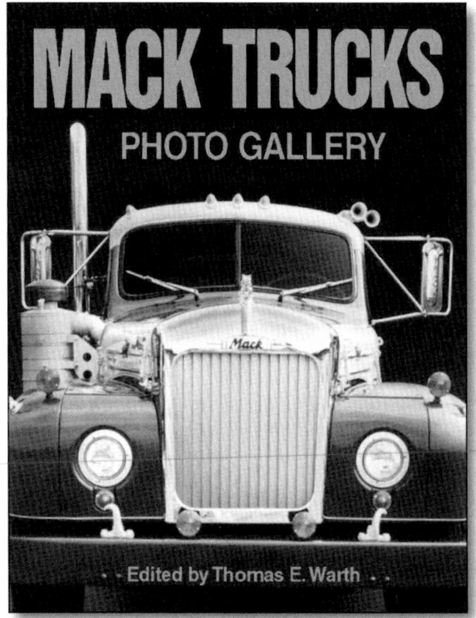

ISBN 1-882256-88-3*

MAXIM
FIRE APPARATUS
PHOTO HISTORY
Howard T. Smith

ISBN 1-58388-111-5